An Introduction To KALMAN FILTERING WITH APPLICATIONS

Kenneth S. Miller, Ph. D.

Riverside Research Institute
New York, NY 10036

and

Donald M. Leskiw, M. S.

Applied Research and Engineering, Inc.
Bedford, MA 01730

ROBERT E. KRIEGER PUBLISHING COMPANY
MALABAR, FLORIDA
1987

Original Edition 1987

Printed and Published by
ROBERT E. KRIEGER PUBLISHING COMPANY, INC.
KRIEGER DRIVE
MALABAR, FL 32950

Library of Congress Cataloging-in-Publication Data

Miller, Kenneth S.
  An introduction to Kalman filtering with applications.

  Bibliography: p.
  Includes index.
  1. Kalman filtering.   2. Prediction theory.
I. Leskiw, Donald M.   II. Title.
QA402.3.M498 1987   515.3′5   85-12606
ISBN 0-89874-824-0

10 9 8 7 6 5 4 3 2

Table of Contents

Chapter IV PRACTICAL ASPECTS OF KALMAN ESTIMATION

Preface

The trajectory of a reentry vehicle is nominally governed by a differential equation. Thus if we know the initial conditions, say at time $t = t_0$, we can predict, with reasonable accuracy, the position of the vehicle at any future time $t = t_1$. Now suppose that, in addition to the differential equation, we are in possession of observations on the position of the vehicle (obtained by a sensor, for example, a radar). Then we have additional information that should lead us to a more accurate estimate of the position of the vehicle at any time $t_1 > t_0$. We recognize that the observations are not one hundred percent accurate; they are corrupted by noise. The basic idea behind Kalman filtering is to combine the *system model* (the differential equation governing the flight path of the vehicle) and the *measurement model* (the observations made on vehicle position) in an optimum fashion in order to obtain the best estimate of the position of the vehicle at any time $t_1 > t_0$. Of course the theory is applicable to almost any differential equation and set of observations. We have used the example of a reentry vehicle for concreteness.

In Chapter I we shall develop, in an efficient and rigorous manner, the basic equations associated with Kalman filtering. While our formulas are based on a linear model, and while there exist many generalizations of the Kalman filter (for example, to nonlinear differential equations) the basic formulas remain the same.

The remaining chapters of the book are concerned with practical aspects of the Kalman theory. We consider a relatively sophisticated dynamical system, namely, the one related to the estimation of the trajectory of a ballistic reentry vehicle. Chapter II discusses the appropriate system model, and Chapter III the associated (radar) measurement model. The final chapter is devoted to certain practical problems that arise in the implementation of the Kalman estimation equations.

Ordinary linear differential equation theory, some probability theory, and elementary mechanics are sufficient prerequisites for an understanding of this book.

April, 1985

K. S. M.
D. M. L.

Chapter I

The Basic Theory

0. INTRODUCTION

Given a dynamical system involving a deterministic vector $\mathbf{x}(t)$ (called the state vector) we wish to determine a best estimate of $\mathbf{x}$ at various times t in the future (or past). Naturally we must identify what information is available with which to construct such estimates, and also define what we mean by a best estimate. The basic ingredients of the Kalman filter which allow us to obtain estimates of the state vector are a stochastic differential system (called the system model) and a set of observations (called the measurement model). The virtue of the Kalman approach is that it is recursive in form. This means that if an estimate $\hat{\mathbf{x}}(t')$ of $\mathbf{x}$ at time $t = t'$ is determined, then all information occurring up to time t' is subsumed in $\hat{\mathbf{x}}(t')$ and its covariance matrix. Thus if we wish to determine a best estimate of $\mathbf{x}$ at a time $t^* > t'$, we need consider only $\hat{\mathbf{x}}(t')$ (and its covariance matrix) and the new information occurring *after t'* and up to t^*. Thus there is no need to retain or store the information that occurred prior to t'.

The set of observations encountered in the measurement model is essentially of the same type that occurs in least squares theory. Thus, in Section 1, we recall a few properties of least squares models, such as the Markov estimate (ME). We then define the best linear unbiased estimate (BLUE) of the state vector and introduce the BL operator. The BL operator leads to a convenient and compact way of defining and determining the

BLUE. Some of the properties and applications of this operator are examined.

A brief review of the theory of ordinary linear differential equations in vector form is a prelude to the discussion of stochastic differential systems (Section 2). Particular care is exercised in defining the equations that arise from taking conditional expectations of the stochastic differential system. We also determine the equations satisfied by the covariance matrix of the solution of the stochastic differential system.

With the above preliminaries in mind we turn to the formal derivation of the Kalman equations. We consider the linear case in order to present the arguments clearly and succinctly, unencumbered by the complications that inevitably arise when nonlinearities are introduced. It should be noted, however, that difficulties arising from nonlinearities have little bearing on the basic Kalman formulation.

A useful variation of the Kalman filter leads to the *backward* Kalman filter. It determines a BLUE $\hat{\mathbf{x}}(t')$ of the state vector at a time $t = t'$ based on all information occurring *after* t'. The arguments in the first paragraph of this section are valid if $t^* < t'$ and $\hat{\mathbf{x}}(t')$ subsumes all information available after t'. In Section 4 the fundamental equations for the backward filter are derived.

Smoothing is defined as the operation of optimally combining two unbiased estimates of the state vector (at a particular time) to obtain an improved estimate. We determine, in Section 5, the basic equations for the smoothed estimate and its covariance matrix.

In Section 6 we demonstrate, in a precisely defined framework, that every least squares problem may be interpreted as a special Kalman filtering problem. Section 7 deals with problems involving control inputs, and shows how such problems may be reduced to cases previously studied. The final section addresses nonlinear Kalman models; and the appropriate form of the Kalman equations is presented.

1. LEAST SQUARES

Linear least squares is a special case of Kalman filtering. Therefore, in this section, we shall review a few elementary properties of least squares theory that will be applicable to the Kalman theory. Additional properties of least squares will be developed as necessary, and its precise relation to Kalman filtering will be presented in Section 6.

In the classical theory one attempts to determine an estimate of a parameter vector $\mathbf{x}$ based on the vector observation

$$\mathbf{z} = H\mathbf{x} + \boldsymbol{v} \ . \tag{1}$$

If $\mathbf{x}$ is a p-dimensional (column) parameter vector, and $\mathbf{z}$ an n-dimensional (column) vector, then H is an $n \times p$ matrix (of known constants). The random vector $\boldsymbol{v}$ is assumed to have mean $\mathbf{0}$ and $n \times n$ positive definite covariance matrix R. (We shall consistently use bold-face characters for vectors.) Naturally, many estimators of $\mathbf{x}$ (based on $\mathbf{z}$) may be found. However, we are generally interested in the *Markov estimate* (ME) $\hat{\mathbf{x}}$ of $\mathbf{x}$. We call $\hat{\mathbf{x}}$ a Markov estimate of $\mathbf{x}$ if each component $\hat{x}_k$ of $\hat{\mathbf{x}}$ is a minimum variance linear unbiased estimator of the corresponding component x_k of $\mathbf{x}$. Under the assumption that $n \geqq p$ and that H has maximal rank p, the familiar arguments (see, for example [37]) show that

$$\hat{\mathbf{x}} = (H'\Sigma^{-1}H)^{-1}H'\Sigma^{-1}\mathbf{z} \tag{2}$$

is the ME of $\mathbf{x}$, and that the covariance matrix P of $\hat{\mathbf{x}}$ is

$$P = (H'\Sigma^{-1}H)^{-1} \ . \tag{3}$$

(The prime on the matrix H in the above two equations indicates transpose. Thus a primed vector is a row vector.)

We shall approach the above problem slightly more generally, and derive some formulas that will be useful in all aspects of our future work. Suppose, then, that $\mathbf{z}_k$, $1 \leqq k \leqq m$, are m linear, but not necessarily unbiased, independent estimates of the parameter vector $\mathbf{x}$. Then $\mathbf{z}_k$ may be written in the form

$$\mathbf{z}_k = H_k\mathbf{x} + \mathbf{a}_k + \boldsymbol{v}_k \ , \ 1 \leqq k \leqq m$$

where $\boldsymbol{v}_k$ has mean zero and positive definite covariance matrix R_k. Thus $\mathscr{E}\mathbf{z}_k = H_k\mathbf{x} + \mathbf{a}_k$. The additive part of the bias may be removed by considering $\hat{\mathbf{x}}_k = \mathbf{z}_k - \mathbf{a}_k$ in place of $\mathbf{z}_k$. We propose to find an estimate of $\mathbf{x}$ based on the $\hat{\mathbf{x}}_k$. To do this we must place some restrictions on the H_k (for example, if all the H_k are zero, our problem would be meaningless). We shall therefore assume that

$$\mathbf{H}' = [H_1' \cdots H_m']$$

has rank p.

Now let us attempt to find matrices $A_1, \cdots, A_m$ such that

$$\hat{\mathbf{x}} = A_1\hat{\mathbf{x}}_1 + \cdots + A_m\hat{\mathbf{x}}_m \tag{4}$$

is an unbiased estimate of $\mathbf{x}$ and such that the trace of the covariance matrix of $\hat{\mathbf{x}}$ is minimized. We shall call such an estimator a ''best linear unbiased estimate'' (BLUE) of $\mathbf{x}$.

If $\hat{\mathbf{x}}$ is to be unbiased, we must have

$$\mathcal{E}\hat{\mathbf{x}} = \left(\sum_{k=1}^{m} A_k H_k \right)\mathbf{x} = \mathbf{x}$$

or

$$I = \sum_{k=1}^{m} A_k H_k \tag{5}$$

where I is the $p \times p$ identity matrix. Now if the H_k are arbitrary, it may not be possible to choose matrices A_k such that (5) holds (suppose all the H_k were equal and of rank less than p). However, under our assumption that $\mathbf{H}$ has rank p, it *is* possible to choose matrices A_k such that (5) holds. The covariance matrix P of $\hat{\mathbf{x}}$ is

$$P = \sum_{k=1}^{m} A_k R_k A_k' \tag{6}$$

and we must minimize tr P subject to the constraint (5). Thus we have a constrained minimization problem. Let Λ be a Lagrange matrix multiplier and consider

$$t = \mathrm{tr}\sum_{k=1}^{m} A_k R_k A_k' - 2\mathrm{tr}\Lambda'\left(\sum_{k=1}^{m} A_k H_k \right) .$$

The derivative of the scalar t with respect to the matrix A_k is (see [37])

$$\frac{\partial t}{\partial A_k} = 2A_k R_k - 2\Lambda H_k' , \quad 1 \leqq k \leqq m .$$

Setting these derivatives equal to the zero matrix leads to

$$A_k = \Lambda H_k' R_k^{-1} , \tag{7}$$

and from (6)

$$P = \Lambda \sum_{k=1}^{m} H_k' A_k' .$$

Equation (5) now implies that

$$P = \Lambda \tag{8}$$

(and hence Λ is symmetric). From (5) and (7)

$$I = \Lambda \sum_{k=1}^{m} H_k' R_k^{-1} H_k$$

and from (8)

$$P^{-1} = \sum_{k=1}^{m} H_k' R_k^{-1} H_k \ . \tag{9}$$

Also from (4) and (7)

$$\hat{\mathbf{x}} = P \sum_{k=1}^{m} H_k' R_k^{-1} \hat{\mathbf{x}}_{\mathbf{k}} \ . \tag{10}$$

Equations (9) and (10) are our desired formulas. If $m = 1$, then (10) becomes identical with the ME of (2).

Actually our BLUE $\hat{\mathbf{x}}$ of $\mathbf{x}$ [given by (10)] is slightly more general than it appears. We recall that in order to determine $\hat{\mathbf{x}}$ we minimized the expected value of

$$(\hat{\mathbf{x}} - \mathbf{x})'(\hat{\mathbf{x}} - \mathbf{x}) \tag{11}$$

subject to the constraint that $\mathcal{E}\hat{\mathbf{x}} = \mathbf{x}$. Now suppose we determine an unbiased estimate $\hat{\boldsymbol{\xi}}$ of $\mathbf{x}$ by minimizing the expectation of the quadratic form

$$(\hat{\mathbf{x}} - \mathbf{x})'\Psi(\hat{\mathbf{x}} - \mathbf{x}) \tag{12}$$

where Ψ is an arbitrary positive definite matrix. We see that (11) is a special case of (12) where Ψ is the identity matrix. The same arguments as used above show that $\hat{\boldsymbol{\xi}} = \hat{\mathbf{x}}$ with probability one. That is, the estimate is *independent* of our choice of Ψ.

As a notational convenience we shall write

$$\hat{\mathbf{x}} = \mathrm{BL}[\hat{\mathbf{x}}_1, \cdots, \hat{\mathbf{x}}_m | \mathbf{x}] \tag{13a}$$

or

$$\hat{\mathbf{x}} = \mathrm{BL}[\Xi_m | \mathbf{x}] \tag{13b}$$

to indicate that $\hat{\mathbf{x}}$ is a BLUE of $\mathbf{x}$ based on the set $\Xi_m = \{\hat{\mathbf{x}}_1, \cdots, \hat{\mathbf{x}}_m\}$. Clearly (13) implies that $\hat{\mathbf{x}}$ is given by (10) and that its inverse covariance matrix is given by (9). The crucial point to remember when manipulating the BL operator is that the random variables appearing in (13) must have an expectation of the form of a matrix times $\mathbf{x}$ [where $\mathbf{x}$ is the parameter vector appearing to the right of the vertical bar in (13)].

We readily see that the BL operator enjoys the following properties:

(i) If $C_1, \cdots, C_m$ are nonsingular matrices, then

$$\mathrm{BL}[C_1\hat{\mathbf{x}}_1, \cdots, C_m\hat{\mathbf{x}}_m|\mathbf{x}] = \mathrm{BL}[\Xi_m|\mathbf{x}] = \hat{\mathbf{x}} \ .$$

(ii) If r is a positive integer less than m, and $\Xi_r = \{\hat{\mathbf{x}}_1, \cdots, \hat{\mathbf{x}}_r\}$, $\Xi_r^* = \{\hat{\mathbf{x}}_{r+1}, \cdots, \hat{\mathbf{x}}_m\}$, then

$$\mathrm{BL}[\mathrm{BL}[\Xi_r|\mathbf{x}], \ \mathrm{BL}[\Xi_r^*|\mathbf{x}]|\mathbf{x}]$$

$$= \mathrm{BL}[\mathrm{BL}[\Xi_r|\mathbf{x}], \ \Xi_r^*|\mathbf{x}]$$

$$= \mathrm{BL}[\Xi_m|\mathbf{x}] = \hat{\mathbf{x}} \ .$$

(iii) Let $\boldsymbol{\xi}_k$ be a mean zero random vector of the same dimension as $\hat{\mathbf{x}}_k$ with positive definite covariance matrix $S_k, 1 \leq k \leq m$. If $\boldsymbol{\xi}_1, \cdots, \boldsymbol{\xi}_m$ and Ξ_m are independent, then

$$\mathrm{BL}[\hat{\mathbf{x}}_1 + \boldsymbol{\xi}_1, \cdots, \hat{\mathbf{x}}_m + \boldsymbol{\xi}_m|\mathbf{x}]$$

and its inverse covariance matrix are given by (10) and (9) respectively with R_k replaced by $R_k + S_k$ and $\hat{\mathbf{x}}_k$ replaced by $\hat{\mathbf{x}}_k + \boldsymbol{\xi}_k$.

(iv) If C is a nonsingular matrix, then

$$\widehat{C\mathbf{x}} = \mathrm{BL}[\Xi_m|C\mathbf{x}] = C \ \mathrm{BL}[\Xi_m|\mathbf{x}] = C\hat{\mathbf{x}} \ .$$

For example, let us prove (iv). From (i)

$$\mathrm{BL}[\hat{\mathbf{x}}_1, \cdots, \hat{\mathbf{x}}_m|C\mathbf{x}] = \mathrm{BL}[C\hat{\mathbf{x}}_1, \cdots, C\hat{\mathbf{x}}_m|C\mathbf{x}] \ .$$

Let $\hat{\mathbf{y}}_k = C\hat{\mathbf{x}}_k$, $\mathbf{y} = C\mathbf{x}$. Then

$$\hat{\mathbf{y}} = \widehat{C\mathbf{x}} = \mathrm{BL}[\hat{\mathbf{y}}_1, \cdots, \hat{\mathbf{y}}_m|\mathbf{y}] \ .$$

Now $\mathcal{E}\hat{\mathbf{y}}_k = J_k\mathbf{y}$ for some J_k. Also

$$\mathcal{E}\hat{\mathbf{y}}_k = \mathcal{E}C\hat{\mathbf{x}}_k = CH_k\mathbf{x} \ .$$

Since $\mathbf{y} = C\mathbf{x}$ we infer

$$J_k = CH_kC^{-1} \ .$$

Also, if we let the covariance matrix of $\hat{\mathbf{y}}_k$ be S_k, then

$$S_k = \mathcal{E}(\hat{\mathbf{y}}_k - \mathcal{E}\hat{\mathbf{y}}_k)(\hat{\mathbf{y}}_k - \mathcal{E}\hat{\mathbf{y}}_k)'$$

$$= \mathcal{E}(C\hat{\mathbf{x}}_k - CH_k\mathbf{x})(C\hat{\mathbf{x}}_k - CH_k\mathbf{x})'$$

$$= CR_kC' \ .$$

Finally, if we let

$$Q^{-1} = \sum_{k=1}^{m} J_k' S_k^{-1} J_k \ ,$$

then

$$Q^{-1} = \sum_{k=1}^{m} (CH_k C^{-1})'(CR_k C')^{-1}(CH_k C^{-1})$$

$$= C'^{-1}\left[\sum_{k=1}^{m} H_k' R_k^{-1} H_k\right] C^{-1}$$

$$= C'^{-1} P^{-1} C^{-1}$$

[see (9)] and $Q = CPC'$. Hence

$$\hat{\mathbf{y}} = \widehat{C\mathbf{x}} = Q \sum_{k=1}^{m} J_k' S_k^{-1} \hat{\mathbf{y}}_k$$

$$= (CPC') \sum_{k=1}^{m} (CH_k C^{-1})'(CR_k C')^{-1}(C\hat{\mathbf{x}}_k)$$

$$= C\left[P \sum_{k=1}^{m} H_k' R_k^{-1} \hat{\mathbf{x}}_k\right]$$

$$= C\hat{\mathbf{x}}$$

[see (10)]. Thus

$$\widehat{C\mathbf{x}} = C \ \mathrm{BL}[\hat{\mathbf{x}}_1, \cdots, \hat{\mathbf{x}}_m | \mathbf{x}] = C\hat{\mathbf{x}}$$

[see (13a)]—which is (iv).

Property (ii) is a fundamental identity which will be useful in future work. For example, suppose $\hat{\mathbf{x}}$ is given by (13) and that an additional independent estimate $\hat{\mathbf{x}}_{m+1}$ of $\mathbf{x}$ is given. Then to determine $\mathrm{BL}[\hat{\mathbf{x}}_1, \cdots, \hat{\mathbf{x}}_{m+1} | \mathbf{x}]$, that is, the BLUE of $\mathbf{x}$ based on the total data set $\hat{\mathbf{x}}_1, \cdots, \hat{\mathbf{x}}_{m+1}$, we need determine only $\mathrm{BL}[\hat{\mathbf{x}}, \hat{\mathbf{x}}_{m+1} | \mathbf{x}]$. From the point of view of computer utilization, there is a tremendous savings in storage.

Explicitly, suppose $\hat{\mathbf{x}}_{m+1}$ is of the same form as $\hat{\mathbf{x}}_k$, $1 \leq k \leq m$. That is,

$$\hat{\mathbf{x}}_{m+1} = H_{m+1}\mathbf{x} + \boldsymbol{v}_{m+1}$$

where $\mathbf{v}_{m+1}$ has mean zero and positive definite covariance matrix R_{m+1}. Then if $\hat{\mathbf{y}} = \mathrm{BL}[\hat{\mathbf{x}}_1, \cdots, \hat{\mathbf{x}}_{m+1} | \mathbf{x}]$, we have

$$\hat{\mathbf{y}} = Q \sum_{k=1}^{m+1} H_k' R_k^{-1} \hat{\mathbf{x}}_k \tag{14}$$

and

$$Q^{-1} = \sum_{k=1}^{m+1} H_k' R_k^{-1} H_k \tag{15}$$

where Q is the covariance matrix of $\hat{\mathbf{y}}$.

Let us now obtain a form for $\hat{\mathbf{y}}$ involving only $\hat{\mathbf{x}}, \hat{\mathbf{x}}_{m+1}$ and H_{m+1}, R_{m+1}, Q and a form for Q involving only P and H_{m+1}, R_{m+1}. But this is trivial. Since $\hat{\mathbf{y}} = \mathrm{BL}[\hat{\mathbf{x}}, \hat{\mathbf{x}}_{m+1} | \mathbf{x}]$ by Property (ii) of the BL operator,

$$Q^{-1} = P^{-1} + H_{m+1}' R_{m+1}^{-1} H_{m+1} \tag{16}$$

and

$$\hat{\mathbf{y}} = Q[P^{-1}\hat{\mathbf{x}} + H_{m+1}' R_{m+1}^{-1} \hat{\mathbf{x}}_{m+1}] \ . \tag{17}$$

Now replace P^{-1} as given by (16) in the above equation to obtain

$$\hat{\mathbf{y}} = (I - GH_{m+1})\hat{\mathbf{x}} + G\hat{\mathbf{x}}_{m+1} \tag{18}$$

where

$$G = QH_{m+1}' R_{m+1}^{-1} \ . \tag{19}$$

Equations (18) and (16) are recursive forms of the least squares estimator and its covariance matrix. The matrix G of (19) is sometimes called the gain matrix.

We conclude this section with a derivation of the BLUE $\hat{\mathbf{x}}$ of the parameter vector $\mathbf{x}$ based on *correlated* estimates. The formulas we derive for $\hat{\mathbf{x}}$ and its covariance matrix P [see (26) and (25)], although quite elegant, are less explicit than the formulas [see (10) and (9)] that we obtained in the independent case. Suppose, then, that as before the linear estimates $\hat{\mathbf{x}}_1, \cdots, \hat{\mathbf{x}}_m$ have means $H_1 \mathbf{x}, \cdots, H_m \mathbf{x}$ respectively. However, let us assume that the $\hat{\mathbf{x}}_k$ are no longer necessarily independent. We shall let $\mathbf{R}$ be the covariance matrix of the partitioned vector

$$\hat{\mathbf{X}}' = \{\hat{\mathbf{x}}_1', \cdots, \hat{\mathbf{x}}_m'\} \ .$$

Then the block diagonal matrices of $\mathbf{R}$ are precisely the covariance matrices R_k of the $\hat{\mathbf{x}}_k$.

As before, we wish to choose matrices $A_1, \cdots, A_m$ such that

$$\hat{\mathbf{x}} = A_1\hat{\mathbf{x}}_1 + \cdots + A_m\hat{\mathbf{x}}_m$$

is a BLUE of $\mathbf{x}$. We may write the above equation succinctly as

$$\hat{\mathbf{x}} = \mathbf{A}\hat{\mathbf{X}} \qquad (20)$$

where

$$\mathbf{A} = [A_1 \cdots A_m]$$

is a partitioned matrix. For $\hat{\mathbf{x}}$ to be unbiased we must have

$$\mathbf{AH} = I \qquad (21)$$

where

$$\mathbf{H}' = [H_1' \cdots H_m'] \ .$$

Now the covariance matrix P of $\hat{\mathbf{x}}$ is given by

$$P = \mathbf{ARA}' \ , \qquad (22)$$

and minimizing

$$t = \operatorname{tr} \mathbf{ARA}' - 2\operatorname{tr}\Lambda'\mathbf{AH}$$

where Λ is a Lagrange matrix multiplier leads to

$$\mathbf{A} = \Lambda\mathbf{H}'\mathbf{R}^{-1} \ . \qquad (23)$$

Thus from (22) and (21)

$$P = \Lambda\mathbf{H}'\mathbf{A}' = \Lambda \ . \qquad (24)$$

But from (21) and (23)

$$I = \Lambda\mathbf{H}'\mathbf{R}^{-1}\mathbf{H}$$

and by (24) we see that

$$P^{-1} = \mathbf{H}'\mathbf{R}^{-1}\mathbf{H} \ . \qquad (25)$$

Also from (20) and (23) and (24)

$$\hat{\mathbf{x}} = P\mathbf{H}'\mathbf{R}^{-1}\hat{\mathbf{X}} \ . \qquad (26)$$

[Compare (26) and (25) with the ME formulas of (2) and (3).] If the $\hat{\mathbf{x}}_1, \cdots, \hat{\mathbf{x}}_m$ are independent, then, of course, (25) and (26) reduce to (9) and (10) respectively.

2. DIFFERENTIAL EQUATIONS

The Kalman model involves a stochastic differential equation and a set of observations. Roughly speaking, every least squares problem is equivalent to a Kalman filter where the observations are of the least squares form discussed in the previous section, and the stochastic differential equation is the trivial equation $\dot{\mathbf{X}} = \mathbf{0}$. Later (see Section 6) we shall give a precise formulation of this qualitative statement. However, for the present, let us review a few properties of stochastic differential equations.

We begin by first examining nonstochastic ordinary linear differential equations. The fundamental existence and uniqueness theorem [34,37] states that if $F(t)$ is a square matrix continuous on a closed finite interval T, and if t_0 is any point in T, and $\mathbf{x}_0$ any constant vector, then there exists a unique continuously differentiable vector $\mathbf{x}(t)$ which satisfies the equation

$$\dot{\mathbf{x}}(t) = F(t)\mathbf{x}(t) \tag{1a}$$

for all $t \in T$ and has the property that

$$\mathbf{x}(t_0) = \mathbf{x}_0. \tag{1b}$$

If the dimension of F is $p \times p$, then there exist p linearly independent vectors $\boldsymbol{\phi}_1(t), \cdots, \boldsymbol{\phi}_p(t)$ which satisfy (1a). For example, let

$$\boldsymbol{\phi}_k(t_0) = \boldsymbol{\delta}_k, \; 1 \leqq k \leqq p$$

where $\boldsymbol{\delta}'_k = \{\delta_{jk}, \cdots, \delta_{pk}\}$. (with δ_{jk} the Kronecker delta). Every solution of (1) may be written as a linear combination of the fundamental system $\boldsymbol{\phi}_1(t), \cdots, \boldsymbol{\phi}_p(t)$.

Let

$$W(t) = [\boldsymbol{\phi}_1(t) \cdots \boldsymbol{\phi}_p(t)]$$

be the Wronskian matrix. Then det $W(t)$ is never zero on T. The one-sided Green's function matrix associated with (1a) is

$$W(t,s) = W(t)W^{-1}(s)$$

and is independent of the choice of a fundamental system. The Green's function matrix enjoys the following properties

$$W(t,t) = I \tag{2a}$$

$$W(t,s)W(s,u) = W(t,u) \tag{2b}$$

$$W(t,s)W(s,t) = I \tag{2c}$$

$$W(t,s) = W^{-1}(s,t) \tag{2d}$$

$$\frac{\partial}{\partial t}W(t,s) = F(t)W(t,s) \ . \tag{2e}$$

Let $\mathbf{v}(t)$ be a vector function continuous on T, and consider the differential system

$$\dot{\mathbf{x}}(t) = F(t)\mathbf{x}(t) + \mathbf{v}(t) \tag{3a}$$

$$\mathbf{x}(t_0) = \mathbf{x}_0 \ , \ t_0 \in T \ . \tag{3b}$$

Then the Green's function matrix may be used to solve (3). Precisely,

$$\mathbf{x}(t) = W(t,t_0)\mathbf{x}_0 + \int_{t_0}^{t} W(t,s)\mathbf{v}(s)ds$$

is the unique solution of (3) for all $t \in T$.

If $F(t)$ is a *constant* matrix, then the one-sided Green's function matrix associated with F depends only on the difference of its arguments. That is

$$W(t,s) = W(t-s)W^{-1}(0) \tag{4}$$

where $W(t)$ is any Wronskian matrix of $\dot{\mathbf{x}}(t) = F\mathbf{x}(t)$. If we choose $W(t)$ to be a Wronskian matrix with the property that $W(0) = I$, then (4) becomes

$$W(t,s) = W(t-s) \ .$$

In fact

$$W(t,s) = e^{(t-s)F} \ .$$

So far we have not mentioned stochastic processes in connection with differential systems. If one or more of the quantities $F(t)$, $\mathbf{v}(t)$, $\mathbf{x}_0$ of (3) is random, then we call (3) a stochastic differential system. In the Kalman theory we need consider only differential systems in which $\mathbf{v}(t)$, or $\mathbf{x}_0$, or both are random. In this connection we have the following important theorem [37]: Let $\{\mathbf{V}(t)|t \in T\}$ be a mean zero mean square continuous process defined on a probability space $\mathbf{P}$. Let $F(t)$ be a square matrix continuous on T, and let $W(t,s)$ be the one-sided Green's function matrix associated with $F(t)$. Let $\hat{\mathbf{x}}_0$ be a random vector defined on $\mathbf{P}$. Then

$$\mathbf{X}(t) = W(t,t_0)\hat{\mathbf{x}}_0 + \int_{t_0}^{t} W(t,s)\mathbf{V}(s)ds \tag{5}$$

is the unique mean square solution of the differential system

$$\dot{\mathbf{X}}(t) = F(t)\mathbf{X}(t) + \mathbf{V}(t) \tag{6a}$$

$$\mathbf{X}(t_0) = \hat{\mathbf{x}}_0 \ , \ t_0 \in T \ . \tag{6b}$$

Note that $\mathbf{X}(t)$ depends on both $\mathbf{V}(t)$ and $\hat{\mathbf{x}}_0$. We shall assume that $\mathbf{V}(t)$ and $\hat{\mathbf{x}}_0$ are independent. Now let us define the expectation of $\mathbf{X}(t)$ conditioned on $\hat{\mathbf{x}}_0$ as

$$\hat{\mathbf{x}}(t) = \mathcal{E}[\mathbf{X}(t)|\hat{\mathbf{x}}_0] \ .$$

That is, the random vector $\hat{\mathbf{x}}(t)$ is a conditional expectation. In particular

$$\hat{\mathbf{x}}(t_0) = \mathcal{E}[\mathbf{X}(t_0)|\hat{\mathbf{x}}_0] = \mathcal{E}[\hat{\mathbf{x}}_0|\hat{\mathbf{x}}_0] = \hat{\mathbf{x}}_0 \ .$$

We also define the expectation of $\mathbf{X}(t)$ conditioned on $\mathbf{V}(t)$ as

$$\mathbf{\Xi}(t) = \mathcal{E}[\mathbf{X}(t)|\mathbf{V}(t)] \ .$$

Thus the random vector $\mathbf{\Xi}(t)$ also is a conditional expectation. In particular

$$\mathbf{\Xi}(t_0) = \mathcal{E}[\mathbf{X}(t_0)|\mathbf{V}(t)] = \mathcal{E}[\hat{\mathbf{x}}_0|\mathbf{V}(t)]$$

$$= \mathcal{E}\hat{\mathbf{x}}_0 = (\text{say}) \ \mathbf{x}_0 \ .$$

Finally, let $\mathbf{x}(t)$ be the unconditional expectation of $\mathbf{X}(t)$, viz.:

$$\mathbf{x}(t) = \mathcal{E}\mathbf{X}(t) = \mathcal{E}[\hat{\mathbf{x}}(t)|\mathbf{V}(t)] = \mathcal{E}[\mathbf{\Xi}(t)|\hat{\mathbf{x}}_0] \ .$$

In particular

$$\mathbf{x}(t_0) = \mathcal{E}\mathbf{X}(t_0) = \mathcal{E}\hat{\mathbf{x}}_0 = \mathbf{x}_0 \ .$$

Let us now interpret (6) in terms of these various expectations. Three cases exist:

Case I $\mathbf{X}(t)$ conditioned on $\hat{\mathbf{x}}_0$

Case II $\mathbf{X}(t)$ conditioned on $\mathbf{V}(t)$

Case III Unconditional expectation of $\mathbf{X}(t)$.

For Case I we have

$$\hat{\mathbf{x}}(t) = W(t,t_0)\hat{\mathbf{x}}_0 \tag{7}$$

from (5) and

$$\dot{\hat{\mathbf{x}}}(t) = \frac{\partial}{\partial t} W(t,t_0)\hat{\mathbf{x}}_0 = F(t)W(t,t_0)\hat{\mathbf{x}}_0 = F(t)\hat{\mathbf{x}}(t)$$

from (2e) and (7). Thus, conditioned on $\hat{\mathbf{x}}_0$, the differential system of (6) becomes

$$\dot{\hat{\mathbf{x}}}(t) = F(t)\hat{\mathbf{x}}(t) \tag{8a}$$

$$\hat{\mathbf{x}}(t_0) = \hat{\mathbf{x}}_0 \ . \tag{8b}$$

For Case II we have

$$\boldsymbol{\Xi}(t) = W(t,t_0)\hat{\mathbf{x}}_0 + \int_{t_0}^{t} W(t,s)\mathbf{V}(s)ds \tag{9}$$

from (5) and

$$\dot{\boldsymbol{\Xi}}(t) = F(t)W(t,t_0)\hat{\mathbf{x}}_0 + \mathbf{V}(t) + F(t)\int_{t_0}^{t} W(t,s)\mathbf{V}(s)ds$$

$$= F(t)\boldsymbol{\Xi}(t) + \mathbf{V}(t)$$

from (2e) and (9). Thus, conditioned on $\mathbf{V}(t)$, the differential system of (6) becomes

$$\dot{\boldsymbol{\Xi}}(t) = F(t)\boldsymbol{\Xi}(t) + \mathbf{V}(t) \tag{10a}$$

$$\boldsymbol{\Xi}(t_0) = \mathbf{x}_0 \text{ with probability } 1 \ . \tag{10b}$$

For Case III we have

$$\mathbf{x}(t) = W(t,t_0)\mathbf{x}_0$$

and

$$\dot{\mathbf{x}}(t) = F(t)\mathbf{x}(t) \tag{11a}$$

$$\mathbf{x}(t_0) = \mathbf{x}_0 \ . \tag{11b}$$

Note that (8a), (10b), (11) are deterministic while (6), (8b), (10a) are random. Thus (2), (8), (10) are stochastic differential systems and (11) is a deterministic ordinary differential system.

Let us return to (6) and assume that $\mathbf{V}(t)$ has cross-covariance matrix

$$S(t,s) = \mathcal{E}\mathbf{V}(t)\mathbf{V}'(s)$$

and that $\hat{\mathbf{x}}_0$ has covariance matrix P_0. Then from (5)

$$\mathcal{E}\mathbf{X}(t) = W(t,t_0)\mathbf{x}_0$$

and the covariance matrix $P(t)$ of $\mathbf{X}(t)$ is

$$P(t) = \mathcal{E}[\mathbf{X}(t) - W(t,t_0)\mathbf{x}_0][\mathbf{X}(t) - W(t,t_0)\mathbf{x}_0]'$$

$$= W(t,t_0)P_0W'(t,t_0) + \int_{t_0}^{t}\int_{t_0}^{t} W(t,s)S(s,\sigma)W'(t,\sigma)\,ds\,d\sigma \ . \tag{12}$$

Sometimes it is convenient to consider a (nonrandom) differential system satisfied by $P(t)$. From (12)

$$\dot{P}(t) = F(t)P(t) + P(t)F'(t) + \int_{t_0}^{t} W(t,s)S(s,t)ds + \int_{t_0}^{t} S(t,s)W'(t,s)ds \quad (13a)$$

where we have used (2e). Equation (13a) together with

$$P(t_0) = P_0 \quad (13b)$$

is the differential system satisfied by $P(t)$.

If

$$S(t,s) = B(t)\Omega(t)B'(t)\delta(t-s)$$

where $\Omega(t)$ is a spectral density matrix and $B(t)$ is a square matrix, then $\{\mathbf{V}(t)|t \in T\}$ is no longer mean square continuous. Nevertheless, one may show that (see [37])

$$P(t) = W(t,t_0)P_0W'(t,t_0) + \int_{t_0}^{t} W(t,s)B(s)\Omega(s)B'(s)W'(t,s)ds$$

if $t > t_0$ and

$$P(t) = W(t,t_0)P_0W'(t,t_0) + \int_{t}^{t_0} W(t,s)B(s)\Omega(s)B'(s)W'(t,s)ds$$

if $t_0 > t$, [see (12)]. Thus from (13a)

$$\dot{P}(t) = F(t)P(t) + P(t)F'(t) + B(t)\Omega(t)B'(t) \quad , \quad t > t_0 \quad (14a)$$

and

$$\dot{P}(t) = F(t)P(t) + P(t)F'(t) - B(t)\Omega(t)B'(t) \quad , \quad t < t_0 \ . \quad (14b)$$

If $\Omega(t)$ is positive definite, then $\Omega^*(t) = B(t)\Omega(t)B'(t)$ is nonnegative definite and one could use Ω^* in place of Ω in the above equations. However, in many practical problems it is convenient to use the representation $B\Omega B'$ rather than Ω^*.

Let $[t_0,t_n]$ be a subinterval of T. Then

$$[t_0,t_n) = \bigcup_{k=1}^{n} [t_{k-1},t_k)$$

where $t_0, \cdots ,t_n$ is a strictly increasing sequence of points. Now suppose that, in addition to the stochastic differential system of (8), we are given independent estimates $\hat{\mathbf{x}}_k$ of $\mathbf{x}$ at times t_k of the form described in the previous section, viz.:

$$\hat{\mathbf{x}}_k = H_k \mathbf{x}(t_k) + \boldsymbol{v}_k \ , \quad 1 \leq k \leq n \ , \tag{15}$$

where H_k is a matrix of constants and $\boldsymbol{v}_k$ has mean zero and positive definite covariance matrix R_k. What we would like to do now is to determine a BLUE of $\mathbf{x}(t_n)$ using the stochastic differential equation *and* the estimates of (15). In the Kalman theory the vector parameter $\mathbf{x}(t)$ is usually referred to as the *state* vector.

Since $\hat{\mathbf{x}}_0$ [see (8b)] is an initial estimate of the state vector $\mathbf{x}$ at time $t = t_0$, we may obtain an estimate $\hat{\mathbf{x}}(t_1^-)$ of $\mathbf{x}$ at time $t = t_1$ by *propagating* the solution throughout the interval $[t_0, t_1)$. From (7)

$$\hat{\mathbf{x}}(t_1^-) = W(t_1, t_0)\hat{\mathbf{x}}_0 \tag{16}$$

and

$$\mathcal{E}\hat{\mathbf{x}}(t_1^-) = W(t_1, t_0)\mathbf{x}_0 = \mathbf{x}(t_1) \ . \tag{17}$$

Then $\hat{\mathbf{x}}(t_1^-)$ and $\hat{\mathbf{x}}_1$ are independent estimators of the state vector $\mathbf{x}$ at time $t = t_1$ and their means are both of the form of a matrix times $\mathbf{x}(t_1)$. The BLUE $\hat{\mathbf{x}}(t_1)$ of $\mathbf{x}(t_1)$ based on $\hat{\mathbf{x}}(t_1^-)$ and $\hat{\mathbf{x}}_1$, that is, an *updated* estimate of $\mathbf{x}(t_1)$, is given by

$$\hat{\mathbf{x}}(t_1) = \mathrm{BL}[\hat{\mathbf{x}}(t_1^-), \hat{\mathbf{x}}_1 | \mathbf{x}(t_1)]$$

$$= \mathrm{BL}[W(t_1, t_0)\hat{\mathbf{x}}_0, \hat{\mathbf{x}}_1 | \mathbf{x}(t_1)]$$

$$= \mathrm{BL}[\hat{\mathbf{x}}_0, \hat{\mathbf{x}}_1 | \mathbf{x}(t_1)] \ . \tag{18}$$

We have used (16) and Property (i) of the BL operator in deriving (18). Now

$$\mathcal{E}\hat{\mathbf{x}}_0 = \mathbf{x}_0 = W^{-1}(t_1, t_0)W(t_1, t_0)\mathbf{x}_0$$

$$= W^{-1}(t_1, t_0)\mathbf{x}(t_1) \tag{19}$$

$$= W(t_0, t_1)\mathbf{x}(t_1)$$

by (17) and (2d). Also

$$\mathcal{E}\hat{\mathbf{x}}_1 = H_1\mathbf{x}(t_1) \ .$$

Thus both random variables appearing in (18) have their means given by a matrix times $\mathbf{x}(t_1)$—as is required in manipulating the BL operator.

Now we may use $\hat{\mathbf{x}}(t_1)$ as an initial condition for the stochastic differential equation (8a) in the subinterval $[t_1, t_2)$ and propagate the solution of the equation throughout $[t_1, t_2)$ to obtain the estimator

$$\hat{\mathbf{x}}(t_2^-) = W(t_2, t_1)\hat{\mathbf{x}}(t_1) \tag{20}$$

of the state vector $\mathbf{x}$ at time $t = t_2$. Taking expectations and using (17) and (2b) we obtain

$$\mathbf{x}(t_2) = W(t_2,t_1)\mathbf{x}(t_1)$$

$$= W(t_2,t_1)[W(t_1,t_0)\mathbf{x}_0] \tag{21}$$

$$= W(t_2,t_0)\mathbf{x}_0 \ .$$

The BLUE of $\hat{\mathbf{x}}(t_2^-)$ and $\hat{\mathbf{x}}_2$, namely

$$\hat{\mathbf{x}}(t_2) = \mathrm{BL}[\hat{\mathbf{x}}(t_2^-),\hat{\mathbf{x}}_2 | \mathbf{x}(t_2)]$$

is an updated estimate of $\mathbf{x}(t_2)$. From (20), (18), Property (iv) of the BL operator, (21) and Property (ii) of the BL operator,

$$\hat{\mathbf{x}}(t_2) = \mathrm{BL}[W(t_2,t_1)\hat{\mathbf{x}}(t_1),\hat{\mathbf{x}}_2 | \mathbf{x}(t_2)]$$

$$= \mathrm{BL}[W(t_2,t_1)\mathrm{BL}[\hat{\mathbf{x}}_0,\hat{\mathbf{x}}_1 | \mathbf{x}(t_1)],\hat{\mathbf{x}}_2 | \mathbf{x}(t_2)]$$

$$= \mathrm{BL}[\mathrm{BL}[\hat{\mathbf{x}}_0,\hat{\mathbf{x}}_1 | W(t_2,t_1)\mathbf{x}(t_1)],\hat{\mathbf{x}}_2 | \mathbf{x}(t_2)]$$

$$= \mathrm{BL}[\mathrm{BL}[\hat{\mathbf{x}}_0,\hat{\mathbf{x}}_1 | \mathbf{x}(t_2)],\hat{\mathbf{x}}_2 | \mathbf{x}(t_2)]$$

$$= \mathrm{BL}[\hat{\mathbf{x}}_0,\hat{\mathbf{x}}_1,\hat{\mathbf{x}}_2 | \mathbf{x}(t_2)] \ . \tag{22}$$

Again we see that

$$\mathcal{E}\hat{\mathbf{x}}_0 = \mathbf{x}_0 = W^{-1}(t_2,t_0)W(t_2,t_0)\mathbf{x}_0$$

$$= W^{-1}(t_2,t_0)\mathbf{x}(t_2)$$

$$= W(t_0,t_2)\mathbf{x}(t_2)$$

and

$$\mathcal{E}\hat{\mathbf{x}}_1 = H_1\mathbf{x}(t_1) = H_1 W^{-1}(t_2,t_1)W(t_2,t_1)\mathbf{x}(t_1)$$

$$= H_1 W(t_1,t_2)\mathbf{x}(t_2)$$

while $\mathcal{E}\hat{\mathbf{x}}_2 = H_2\mathbf{x}(t_2)$. Thus, as required, the expectation of all three random variables occurring in (22) is of the form of a matrix times $\mathbf{x}(t_2)$.

Inductively we see that

$$\hat{\mathbf{x}}(t_n) = \mathrm{BL}[\hat{\mathbf{x}}_0,\hat{\mathbf{x}}_1, \cdots , \hat{\mathbf{x}}_n | \mathbf{x}(t_n)]$$

where

$$\mathcal{E}\hat{\mathbf{x}}_0 = W(t_0,t_n)\mathbf{x}(t_n)$$

and

$$\mathcal{E}\hat{\mathbf{x}}_k = H_k W(t_k,t_n)\mathbf{x}(t_n) \ , \ 1 \leqq k \leqq n \ .$$

Thus by (10) of Section 1,

$$\hat{\mathbf{x}}(t_n) = P_n \left[W'(t_0,t_n)P_0^{-1}\hat{\mathbf{x}}_0 + \sum_{k=1}^{n} W'(t_k,t_n)H_k'R_k^{-1}\hat{\mathbf{x}}_k \right] \tag{23}$$

where P_n is the covariance matrix of $\hat{\mathbf{x}}(t_n)$, and by (9) of Section 1,

$$P_n^{-1} = W'(t_0,t_n)P_0^{-1}W(t_0,t_n) + \sum_{k=1}^{n} W'(t_k,t_n)H_k'R_k^{-1}H_kW(t_k,t_n) \ . \tag{24}$$

It does not seem possible to derive simple formulas analogous to (23) and (24) for the stochastic differential system of (6).

3. THE FORMAL EQUATIONS

To fix our ideas, suppose we have a dynamical system governed by a differential system. For example, the trajectory of a vehicle in flight is governed by a differential system. *If* the equations exactly represent the physical phenomenon, and *if* the initial conditions were known exactly, then the position of the vehicle at any point in time could be obtained from a solution of the differential system. However, in all practical problems, neither of these conditions is true. Thus we must deal with a stochastic differential system and must be satisfied to make probabilistic statements regarding the true location of the vehicle. Suppose, however, that in addition to the differential system we have one or more sensors which measure the coordinates (or some function or functional of the coordinates) of the vehicle in flight. Of course these measurements too are subject to random errors. However, armed with the stochastic differential equation *and* the outputs of the sensors, presumably a better estimate of the position of the vehicle is possible. This is the problem solved by the Kalman filter. It combines, in an optimum fashion, the solution of the differential equation and the independent observations.

Let us now precisely formulate the problem. We assume that we are given a deterministic differential system

$$\dot{\mathbf{x}}(t) = F(t)\mathbf{x}(t) \tag{1a}$$

$$\mathbf{x}(t_0) = \mathbf{x}_0 \tag{1b}$$

which ideally represents the physical phenomenon. The vector $\mathbf{x}(t)$ is called the *state vector*. However, as we remarked above, we must deal with a stochastic differential system,

$$\dot{\mathbf{X}}(t) = F(t)\mathbf{X}(t) + \mathbf{V}(t) \qquad (2a)$$

$$\mathbf{X}(t_0) = \hat{\mathbf{x}}_0 \qquad (2b)$$

where the forcing function $\mathbf{V}(t)$ and the initial condition $\hat{\mathbf{x}}_0$ are random. In the Kalman theory (2) is referred to as the *system model* and $\mathbf{V}(t)$ is called the *system noise*. It is assumed that $\mathbf{V}(t)$ has mean zero. Sometimes we shall assume that $\mathbf{V}(t)$ has cross-covariance matrix $S(t,s) = \mathcal{E}\mathbf{V}(t)\mathbf{V}'(s)$, and sometimes we shall assume that $\mathbf{V}(t)$ is "white noise" so that

$$S(t,s) = B(t)\Omega(t)B'(t)\delta(t-s)$$

where $\Omega(t)$ is a (positive definite) spectral density matrix and $B(t)$ is a square matrix. We also assume that $\hat{\mathbf{x}}_0$ is an unbiased estimate of $\mathbf{x}(t_0)$ with positive definite covariance matrix P_0.

We now postulate the existence of a *measurement model* of the form

$$\mathbf{Z}_k = H_k\mathbf{x}(t_k) + \boldsymbol{v}_k \ , \quad k = 1,2, \cdots$$

where

$$t_0 < t_1 < \cdots .$$

We call $\mathbf{Z}_k$ a *measurement* or an *observation*. The *measurement noise* $\boldsymbol{v}_k$ will be assumed to have mean zero and positive definite covariance matrix $R_k = \mathcal{E}\boldsymbol{v}_k\boldsymbol{v}_k'$. The random initial condition $\hat{\mathbf{x}}_0$, the stochastic process $\mathbf{V}(t)$, and the measurement noises $\boldsymbol{v}_1, \boldsymbol{v}_2, \cdots$ are all assumed to be independent.

Our fundamental problem now may be formulated simply as follows: Using the system and measurement models determine the BLUE of the state vector $\mathbf{x}$ at time $t = t_k$ for $k = 1,2, \cdots$. Many of the mathematical techniques we need to derive the formal Kalman equations already have been employed in the previous sections. One of the great virtues of the Kalman filter is that the formulas we obtain are in *recursive* form. That is, if we obtain a BLUE $\mathbf{X}(t_k)$ of $\mathbf{x}$ at time t_k, then to derive the BLUE $\mathbf{X}(t_{k+1})$ of $\mathbf{x}$ at time t_{k+1} using the measurement $\mathbf{Z}_{k+1}$ we need know only $\mathbf{X}(t_k)$ and its covariance matrix. That is, there is no need to store the previous estimates $\mathbf{X}(t_1), \cdots ,\mathbf{X}(t_{k-1})$ or the previous measurements $\mathbf{Z}_1, \cdots ,\mathbf{Z}_k$ or any of their covariance matrices. All the information we need is contained in $\mathbf{X}(t_k)$, $\mathbf{Z}_{k+1}$ and their covariance matrices.

We proceed with the formal derivation. Given the initial estimate $\hat{\mathbf{x}}_0$ of $\mathbf{x}(t_0)$ we use (2) to propagate the solution throughout the interval $[t_0,t_1)$. If $W(t,s)$ is the one-sided Green's function matrix associated with $F(t)$, then

$$\mathbf{X}(t_1^-) = W(t_1,t_0)\hat{\mathbf{x}}_0 + \int_{t_0}^{t_1} W(t_1,s)\mathbf{V}(s)ds \qquad (3)$$

is an unbiased estimate of the state vector $\mathbf{x}$ at time $t = t_1$. The updated estimate $\mathbf{X}(t_1)$ using the first measurement $\mathbf{Z}_1$ is

$$\mathbf{X}(t_1) = BL[\mathbf{X}(t_1^-), \mathbf{Z}_1 | \mathbf{x}(t_1)]$$

and from Section 1,

$$\mathbf{X}(t_1) = P(t_1)[P^{-1}(t_1^-)\mathbf{X}(t_1^-) + H_1'R_1^{-1}\mathbf{Z}_1] \qquad (4)$$

and

$$P^{-1}(t_1) = P^{-1}(t_1^-) + H_1'R_1^{-1}H_1 \qquad (5)$$

where $P(t_1^-)$ is the covariance matrix of $\mathbf{X}(t_1^-)$ and $P(t_1)$ is the covariance matrix of $\mathbf{X}(t_1)$.

If we set

$$K_1 = P(t_1)H_1'R_1^{-1} \qquad (6)$$

then (4) may be written in the form

$$\mathbf{X}(t_1) = (I - K_1 H_1)\mathbf{X}(t_1^-) + K_1 \mathbf{Z}_1 \qquad (7)$$

since from (5) and (6)

$$I = P(t_1)P^{-1}(t_1^-) + P(t_1)H_1'R_1^{-1}H_1$$

$$= P(t_1)P^{-1}(t_1^-) + K_1 H_1 \ . \qquad (8)$$

The matrix K_1 is called the *Kalman gain matrix*.

Now let $\mathbf{X}(t_1)$ be the initial condition for the system model equation of (2a) in the interval $[t_1,t_2)$. Then we may propagate the solution throughout $[t_1,t_2)$ to obtain

$$\mathbf{X}(t_2^-) = W(t_2,t_1)\mathbf{X}(t_1) + \int_{t_1}^{t_2} W(t_2,s)\mathbf{V}(s)ds$$

as an estimate of $\mathbf{x}(t_2)$. The updated estimate is

$$\mathbf{X}(t_2) = BL[\mathbf{X}(t_2^-), \mathbf{Z}_2 | \mathbf{x}(t_2)]$$

$$= P(t_2)[P^{-1}(t_2^-)\mathbf{X}(t_2^-) + H_2'R_2^{-1}\mathbf{Z}_2]$$

$$= (I - K_2 H_2)\mathbf{X}(t_2^-) + K_2 \mathbf{Z}_2$$

where $P(t_2^-)$ and $P(t_2)$ are the covariance matrices, respectively, of $\mathbf{X}(t_2^-)$ and $\mathbf{X}(t_2)$, while

$$K_2 = P(t_2)H_2'R_2^{-1}$$

is the Kalman gain matrix.

Inductively, and in obvious notation

$$K_k = P(t_k)H_k'R_k^{-1} \ , \ k=1,2,\cdots \tag{9}$$

is the Kalman gain matrix,

$$\mathbf{X}(t_1^-) = W(t_1,t_0)\hat{\mathbf{x}}_0 + \int_{t_0}^{t_1} W(t_1,s)\mathbf{V}(s)ds$$

$$\mathbf{X}(t_k^-) = W(t_k,t_{k-1})\mathbf{X}(t_{k-1}) + \int_{t_{k-1}}^{t_k} W(t_k,s)\mathbf{V}(s)ds \ , \ k=2,3,\cdots$$

are the propagated estimates, while the updated estimate $\mathbf{X}(t_k)$, that is, the BLUE of $\mathbf{X}(t_k^-)$ and $\mathbf{Z}_k$ is given by

$$\mathbf{X}(t_k) = (I - K_kH_k)\mathbf{X}(t_k^-) + K_k\mathbf{Z}_k \ , \ k=1,2,\cdots \ .$$

The covariance matrix $P(t_k^-)$ of $\mathbf{X}(t_k^-)$ is

$$P(t_1^-) = \mathcal{E}[\mathbf{X}(t_1^-) - \mathbf{x}(t_1)][\mathbf{X}(t_1^-) - \mathbf{x}(t_1)]'$$

$$= W(t_1,t_0)P_0W'(t_1,t_0) + \int_{t_0}^{t_1}\int_{t_0}^{t_1} W(t_1,s)S(s,\sigma)W'(t_1,\sigma)dsd\sigma$$

$$\tag{10a}$$

where $P_0 = \mathcal{E}(\hat{\mathbf{x}}_0 - \mathbf{x}_0)(\hat{\mathbf{x}}_0 - \mathbf{x}_0)'$ [see (12) of Section 2] and

$$P(t_k^-) = \mathcal{E}[\mathbf{X}(t_k^-) - \mathbf{x}(t_k)][\mathbf{X}(t_k^-) - \mathbf{x}(t_k)]'$$

$$= W(t_k,t_{k-1})P(t_{k-1})W'(t_k,t_{k-1}) \tag{10b}$$

$$+ \int_{t_{k-1}}^{t_k}\int_{t_{k-1}}^{t_k} W(t_k,s)S(s,\sigma)W'(t_k,\sigma)dsd\sigma \ ,$$

$$k=2,3,\cdots \ .$$

The covariance matrix $P(t_k)$ of $\mathbf{X}(t_k)$ is given by

$$P^{-1}(t_k) = P^{-1}(t_k^-) + H_k'R_k^{-1}H_k \ , \ k=1,2,\cdots \ . \tag{11}$$

If we consider the expectation of (2) conditioned on $\hat{\mathbf{x}}_0$, then (2) becomes

$$\dot{\hat{\mathbf{x}}}(t) = F(t)\hat{\mathbf{x}}(t) \tag{12a}$$

$$\hat{\mathbf{x}}(t_0) = \hat{\mathbf{x}}_0 \tag{12b}$$

[see (8) of Section 2]. Now let

$$\hat{\mathbf{x}}(t_k) = \text{BL}[\hat{\mathbf{x}}(t_k^-),\mathbf{Z}_k\,|\,\mathbf{x}(t_k)].$$

Then we arrive at the equations

$$\hat{\mathbf{x}}(t_k^-) = W(t_k, t_{k-1})\hat{\mathbf{x}}(t_{k-1})$$

$$\hat{\mathbf{x}}(t_k) = (I - K_k H_k)\hat{\mathbf{x}}(t_k^-) + K_k \mathbf{Z}_k \tag{13}$$

where

$$K_k = P(t_k)H_k'R_k^{-1} \, , \, k = 1, 2, \cdots \, .$$

For this reason (12a) is called the *state estimate propagation equation*. It should be noted, however, that $P(t_k^-)$ and $P(t_k)$ which appear explicitly and implicitly in (13) are still given by (10) and (11). We shall henceforth replace the notation $\mathbf{X}(t_k^-)$ by $\hat{\mathbf{x}}(t_k^-)$ and $\mathbf{X}(t_k)$ by $\hat{\mathbf{x}}(t_k)$. [See Property (iii) of the BL operator.]

To summarize, then, the Kalman model, in its most commonly used form may be described as follows:

Let $\mathbf{x}(t)$ be a state vector which satisfies the deterministic ordinary differential system

$$\dot{\mathbf{x}}(t) = F(t)\mathbf{x}(t)$$

$$\mathbf{x}(t_0) = \mathbf{x}_0 \, .$$

Then in the Kalman theory we have:

System model

$$\dot{\mathbf{X}}(t) = F(t)\mathbf{X}(t) + \mathbf{V}(t)$$

$$\mathbf{X}(t_0) = \hat{\mathbf{x}}_0$$

where $\mathbf{V}(t)$ has mean zero and cross-covariance matrix $S(t,s) = B(t)\Omega(t)B'(t)\delta(t-s)$ where $\Omega(t)$ is a positive definite spectral density matrix and $B(t)$ is a square matrix. The random initial condition $\hat{\mathbf{x}}_0$ has mean $\mathbf{x}_0$ and positive definite covariance matrix P_0.

Measurement model

$$\mathbf{Z}_k = H_k\mathbf{x}(t_k) + \mathbf{\nu}_k \, , \, k = 1, 2, \cdots$$

where $\mathbf{\nu}_k$ has mean zero and positive definite covariance matrix R_k.

It is assumed that $\mathbf{V}(t)$, $\hat{\mathbf{x}}_0$, and $\mathbf{\nu}_k$, $k = 1, 2, \cdots$ are independent.

State estimate propagation

$$\dot{\hat{\mathbf{x}}}(t) = F(t)\hat{\mathbf{x}}(t) \, , \, t_0 \leqq t < t_1$$

$$\hat{\mathbf{x}}(t_0) = \hat{\mathbf{x}}_0$$

and

$$\dot{\hat{\mathbf{x}}}(t) = F(t)\hat{\mathbf{x}}(t)$$

has initial condition $\hat{\mathbf{x}}(t_k)$ in $[t_k, t_{k+1})$, $k = 1, 2, \cdots$.

Error covariance propagation

$$\dot{P}(t) = F(t)P(t) + P(t)F'(t) + B(t)\Omega(t)B'(t) \; , \; t_0 \leqq t < t_1$$

$$P(t_0) = P_0$$

and

$$\dot{P}(t) = F(t)P(t) + P(t)F'(t) + B(t)\Omega(t)B'(t)$$

has initial condition $P(t_k)$ in $[t_k, t_{k+1})$, $k = 1, 2, \cdots$.

State estimate update

$$\hat{\mathbf{x}}(t_k) = (I - K_k H_k)\hat{\mathbf{x}}(t_k^-) + K_k \mathbf{Z}_k$$

$$= P(t_k)[P^{-1}(t_k^-)\hat{\mathbf{x}}(t_k^-) + H_k'R_k^{-1}\mathbf{Z}_k] \; , \; k = 1, 2, \cdots$$.

Error covariance update

$$P(t_k) = [P^{-1}(t_k^-) + H_k'R_k^{-1}H_k]^{-1}$$

$$= (I - K_k H_k)P(t_k^-) \; , \; k = 1, 2, \cdots$$.

Gain matrix

$$K_k = P(t_k)H_k'R_k^{-1} \; , \; k = 1, 2, \cdots$$.

An alternate form for the gain matrix is given by

$$K_k = P(t_k^-)H_k'[H_k P(t_k^-)H_k' + R_k]^{-1} \; . \tag{14}$$

To prove (14) we start with the identity

$$I = P(t_k)P^{-1}(t_k^-) + K_k H_k$$

[see (8)] and post-multiply by $P(t_k^-)H_k'$ to obtain

$$P(t_k^-)H_k' = P(t_k)H_k' + K_k H_k P(t_k^-)H_k'. \tag{15}$$

From (9) we have $P(t_k)H_k' = K_k R_k$. Using this result in (15) yields

$$P(t_k^-)H_k' = K_k R_k + K_k H_k P(t_k^-)H_k'$$

$$= K_k[R_k + H_k P(t_k^-)H_k']$$

—which immediately reduces to (14).

Also, from (9) and (11),

$$K_k = [P^{-1}(t_k^-) + H_k' R_k^{-1} H_k] H_k' R_k^{-1} \ . \tag{16}$$

We see that (14) expresses K_k in terms of $P(t_k^-)$ and R_k only, while (16) expresses K_k in terms of $P^{-1}(t_k^-)$ and R_k^{-1} only.

Let us consider now an application of some of the above formulas. Suppose, then, that $x(t)$ is a scalar state vector which satisfies the deterministic ordinary differential system

$$\dot{x}(t) = ax(t) \tag{17a}$$

$$x(t_0) = x_0 \tag{17b}$$

where a is a constant. We shall assume that the system noise is zero and that the system model is

$$\dot{X}(t) = aX(t) \tag{18a}$$

$$X(t_0) = \hat{x}_0 \tag{18b}$$

where $\mathcal{E}\hat{x}_0 = x_0$ and Var $\hat{x}_0 = P_0$. As our measurement model we take

$$Z_k = x(t_k) + v_k \ , \ k = 1,2, \cdots$$

where $\mathcal{E}v_k = 0$ and $\mathcal{E}v_k^2 = R$ (independent of k).

For this problem we see that the one-sided Green's function $W(t,s)$ is given by

$$W(t,s) = e^{a(t-s)} \ .$$

Now let

$$\Delta t = t_k - t_{k-1}$$

independent of k, and for convenience introduce the notation

$$E = e^{a\Delta t} \ . \tag{19}$$

Then an application of the Kalman filter equations yields

$$\hat{x}(t_0) = \hat{x}_0$$

$$\hat{x}(t_1^-) = E\hat{x}_0$$

$$\hat{x}(t_k^-) = E\hat{x}(t_{k-1}) \ , \ k = 2,3, \cdots$$

$$\hat{x}(t_k) = P(t_k)[P^{-1}(t_k^-)\hat{x}(t_k^-) + R^{-1}Z_k] \ , \ k = 1,2, \cdots$$

$$P(t_0) = P_0$$

$$P(t_1^-) = E^2 P_0$$

$$P(t_k^-) = E^2 P(t_{k-1}) \ , \ k = 2, 3, \cdots$$

$$P^{-1}(t_k) = P^{-1}(t_k^-) + R^{-1} \ , \ k = 1, 2, \cdots \ .$$

In this simple case we may solve, inductively, the above recursion relations to obtain

$$\hat{x}(t_n) = P(t_n) E^{-n} \left[P_0^{-1} \hat{x}_0 + R^{-1} \sum_{\alpha=1}^{n} E^\alpha Z_\alpha \right] \ , \ n = 1, 2, \cdots \tag{20}$$

and

$$P^{-1}(t_n) = E^{-2n} \left[P_0^{-1} + R^{-1} \sum_{\alpha=1}^{n} E^{2\alpha} \right] \ , \ n = 1, 2, \cdots \ . \tag{21}$$

[See (23) and (24) of Section 2.]

Now it often happens that we have no information regarding the initial estimate $\hat{x}_0$. Thus its variance must be assumed to be unbounded. If $P_0^{-1} = 0$, then we see from (20) that the value of $\hat{x}_0$ is immaterial and

$$\hat{x}(t_n) = \frac{E^n \sum_{\alpha=1}^{n} E^\alpha Z_\alpha}{\sum_{\alpha=1}^{n} E^{2\alpha}} \ , \ n = 1, 2, \cdots \tag{22}$$

while

$$P(t_n) = \frac{R}{E^{-2n} \sum_{\alpha=1}^{n} E^{2\alpha}} \ , \ n = 1, 2, \cdots \ . \tag{23}$$

If, furthermore, $a = 0$, then

$$\hat{x}(t_n) = \frac{1}{n} \sum_{\alpha=1}^{n} Z_\alpha \ , \ n = 1, 2, \cdots \tag{24}$$

is just the average of the observations (sample means) and

$$P(t_n) = \frac{1}{n} R \ , \ n = 1, 2, \cdots \ . \tag{25}$$

4. THE BACKWARD FILTER

In the previous section we assumed that $t_1, t_2, \cdots$ was a strictly increasing sequence of time points at which measurements $\mathbf{Z}_1, \mathbf{Z}_2, \cdots$ on functions of the state vector $\mathbf{x}$ were made. Then with the aid of the stochastic differential system we computed the unbiased estimates $\hat{\mathbf{x}}(t_1^-)$, $\hat{\mathbf{x}}(t_2^-)$, $\cdots$ of the state vector and their respective updates $\hat{\mathbf{x}}(t_1)$, $\hat{\mathbf{x}}(t_2)$, $\cdots$ as well as the associated covariance matrices. Now suppose that an initial condition is given at time $t = t_{N+1}$ and we wish to construct estimates of $\mathbf{x}$ at the times t_N, t_{N-1}, $\cdots$ based on the observations $\mathbf{Z}_N$, $\mathbf{Z}_{N-1}$, $\cdots$. In this section we shall construct such a filter. It is called the *backward* filter to distinguish it from the Kalman model described in the previous section. When both filters are used in the same problem, we use the adjective *forward* to describe the Kalman filter of Section 3.

Suppose then that $t_0, t_1, \cdots, t_N, t_{N+1}$ is a strictly increasing sequence of time points, and that,

$$\dot{\mathbf{y}}(t) = F(t)\mathbf{y}(t) \tag{1a}$$

$$\mathbf{y}(t_{N+1}) = \mathbf{y}_{N+1} \tag{1b}$$

is a deterministic differential system. Then the system model is

$$\dot{\mathbf{Y}}(t) = F(t)\mathbf{Y}(t) + \mathbf{V}(t) \tag{2a}$$

$$\mathbf{Y}(t_{N+1}) = \hat{\mathbf{y}}_{N+1} \tag{2b}$$

where $\hat{\mathbf{y}}_{N+1}$ has mean $\mathbf{x}(t_{N+1})$ and positive definite covariance matrix Q_{N+1}. The system noise is assumed to have mean zero and cross-covariance matrix

$$S(t,s) = \mathcal{E}\mathbf{V}(t)\mathbf{V}'(s) = B(t)\Omega(t)B'(t)\delta(t-s)$$

where $\Omega(t)$ is a positive definite spectral density matrix and $B(t)$ is a square matrix. The measurement model is

$$\mathbf{Z}_k = H_k\mathbf{x}(t_k) + \boldsymbol{v}_k \ , \quad k = 1, 2, \cdots, N$$

where $\mathcal{E}\boldsymbol{v}_k = \mathbf{0}$ and $\mathcal{E}\boldsymbol{v}_k\boldsymbol{v}_k' = R_k$ (positive definite). The random vectors $\hat{\mathbf{y}}_{N+1}$, $\mathbf{V}(t)$, $\boldsymbol{v}_1$, $\cdots$, $\boldsymbol{v}_N$ are all assumed to be independent . Thus (1) and (2), except for the initial conditions are identical with (1) and (2) of Section 3. That is, (1a) and (2a) are identical with (1a) and (2a) respectively of Section 3. The measurement model is identical in both cases.

If, as before, $W(t,s)$ is the one-sided Green's function matrix associated with $F(t)$, then in the interval $(t_N, t_{N+1}]$,

$$\mathbf{Y}(t_N^+) = W(t_N, t_{N+1})\hat{\mathbf{y}}_{N+1} + \int_{t_{N+1}}^{t_N} W(t_N, s)\mathbf{V}(s)ds \tag{3}$$

is an unbiased estimate of the state vector $\mathbf{x}$ at time $t = t_N$. The updated estimate $\mathbf{Y}(t_N)$ using $\mathbf{Z}_N$ is

$$\mathbf{Y}(t_N) = \mathrm{BL}[\mathbf{Y}(t_N^+), \mathbf{Z}_N | \mathbf{x}(t_N)]$$

$$= Q(t_N)[Q^{-1}(t_N^+)\mathbf{Y}(t_N^+) + H_N'R_N^{-1}\mathbf{Z}_N] \ , \tag{4}$$

and

$$Q^{-1}(t_N) = Q^{-1}(t_N^+) + H_N'R_N^{-1}H_N \tag{5}$$

where $Q(t_N^+)$ is the covariance matrix of $\mathbf{Y}(t_N^+)$ and $Q(t_N)$ is the covariance matrix of $\mathbf{Y}(t_N)$. If we let

$$L_N = Q(t_N)H_N'R_N^{-1} \tag{6}$$

be the Kalman gain matrix, then (4) may be written in the form

$$\mathbf{Y}(t_N) = (I - L_N H_N)\mathbf{Y}(t_N^+) + L_N\mathbf{Z}_N \ . \tag{7}$$

Now let $\mathbf{Y}(t_N)$ be the initial condition for the system model equation of (2a) in the interval $(t_{N-1}, t_N]$. Then we may propagate the solution throughout $(t_{N-1}, t_N]$ to obtain

$$\mathbf{Y}(t_{N-1}^+) = W(t_{N-1}, t_N)\mathbf{Y}(t_N) + \int_{t_N}^{t_{N-1}} W(t_{N-1}, s)\mathbf{V}(s)ds$$

as an unbiased estimate of $\mathbf{x}(t_{N-1})$. The updated estimate is

$$\mathbf{Y}(t_{N-1}) = \mathrm{BL}[\mathbf{Y}(t_{N-1}^+), \mathbf{Z}_{N-1} | \mathbf{x}(t_{N-1})]$$

$$= (I - L_{N-1}H_{N-1})\mathbf{Y}(t_{N-1}^+) + L_{N-1}\mathbf{Z}_{N-1}$$

where $Q(t_{N-1}^+)$ and $Q(t_{N-1})$ are the covariance matrices, respectively, of $\mathbf{Y}(t_{N-1}^+)$ and $\mathbf{Y}(t_{N-1})$, and

$$L_{N-1} = Q(t_{N-1})H_{N-1}'R_{N-1}^{-1}$$

is the Kalman gain matrix. The pattern of propagation and updating is now clear.

We turn to the covariance matrices. By definition

$$Q(t_N^+) = \mathcal{E}[\mathbf{Y}(t_N^+) - \mathbf{x}(t_N)][\mathbf{Y}(t_N^+) - \mathbf{x}(t_N)]'$$

$$= W(t_N, t_{N+1})Q_{N+1}W'(t_N, t_{N+1})$$

$$+ \int_{t_{N+1}}^{t_N}\int_{t_{N+1}}^{t_N} W(t_N, s)B(s)\Omega(s)B'(s)\delta(s - \sigma)W'(t_N, \sigma)ds d\sigma$$

$$= W(t_N, t_{N+1})Q_{N+1}W'(t_N, t_{N+1})$$

$$- \int_{t_{N+1}}^{t_N} W(t_N, s)B(s)\Omega(s)B'(s)W'(t_N, s)ds$$

(see Section 2). The updated covariance matrix $Q(t_N)$ is defined by (5). Continuing,

$$Q(t_{N-1}^+) = \mathscr{E}[\mathbf{Y}(t_{N-1}^+) - \mathbf{x}(t_{N-1})][\mathbf{Y}(t_{N-1}^+) - \mathbf{x}(t_{N-1})]'$$

$$= W(t_{N-1},t_N)Q(t_N)W'(t_{N-1},t_N)$$

$$- \int_{t_N}^{t_{N-1}} W(t_{N-1},s)B(s)\Omega(s)B'(s)W'(t_{N-1},s)ds \ .$$

As we did for the forward filter of Section 3 we may summarize the description of the backward filter as follows. Let

$$\dot{\mathbf{y}}(t) = F(t)\mathbf{y}(t)$$

$$\mathbf{y}(t_{N+1}) = \mathbf{y}_{N+1}$$

be the ordinary differential system satisfied by the state vector $\mathbf{x}(t)$ with $\mathbf{y}_{N+1} \equiv \mathbf{x}_{N+1}$. Then in the Kalman framework, the backward filter is described by:

System model

$$\dot{\mathbf{Y}}(t) = F(t)\mathbf{Y}(t) + \mathbf{V}(t)$$

$$\mathbf{Y}(t_{N+1}) = \hat{\mathbf{y}}_{N+1}$$

where $\mathbf{V}(t)$ has mean zero and cross-covariance matrix $Q(t,s) = B(t)\Omega(t)B'(t)\delta(t-s)$ where $\Omega(t)$ is a positive definite spectral density matrix and $B(t)$ is a square matrix. The random initial condition $\hat{\mathbf{y}}_{N+1}$ has mean $\mathbf{x}(t_{N+1})$ and positive definite covariance matrix Q_{N+1}.

Measurement model

$$\mathbf{Z}_k = H_K\mathbf{x}(t_k) + \boldsymbol{v}_k \ , \ k = 1, 2, \cdots, N$$

where $\boldsymbol{v}_k$ has mean zero and positive definite covariance matrix R_k.

It is assumed that $\mathbf{V}(t)$, $\hat{\mathbf{y}}_{N+1}$, and $\boldsymbol{v}_k, k = 1, 2, \cdots, N$ are independent.

State estimate propagation

$$\dot{\hat{\mathbf{y}}}(t) = F(t)\hat{\mathbf{y}}(t) \ , \ t_N < t \leq t_{N+1}$$

$$\hat{\mathbf{y}}(t_{N+1}) = \hat{\mathbf{y}}_{N+1}$$

and

$$\dot{\hat{\mathbf{y}}}(t) = F(t)\hat{\mathbf{y}}(t)$$

has initial condition $\hat{\mathbf{y}}(t_{k+1})$ in $(t_k, t_{k+1}]$, $k = 1, 2, \cdots, N-1$.

Error covariance propagation

$$\dot{Q}(t) = F(t)Q(t) + Q(t)F'(t) - B(t)\Omega(t)B'(t) \ , \ t_N < t \leq t_{N+1}$$

$$Q(t_{N+1}) = Q_{N+1}$$

and

$$\dot{Q}(t) = F(t)Q(t) + Q(t)F'(t) - B(t)\Omega(t)B'(t)$$

has initial condition $Q(t_{k+1})$ in $(t_k, t_{k+1}]$, $k = 1, 2, \cdots, N-1$.

State estimate update

$$\hat{\mathbf{y}}(t_k) = (I - L_k H_k)\hat{\mathbf{y}}(t_k^+) + L_k \mathbf{Z}_k$$

$$= Q(t_k)[Q^{-1}(t_k^+)\hat{\mathbf{y}}(t_k^+) + H_k' R_k^{-1} \mathbf{Z}_k] \ , \ k = 1, 2, \cdots, N$$

Error covariance update

$$Q(t_k) = [Q^{-1}(t_k^+) + H_k' R_k^{-1} H_k]^{-1} \ , \ k = 1, 2, \cdots, N$$

Gain matrix

$$L_k = Q(t_k)H_k' R_k^{-1} \ , \ k = 1, 2, \cdots, N \ .$$

There is no harm in letting $t_{N+1} = t_N^+$, $Q_{N+1} = Q(t_N^+)$ since $\mathbf{Z}_N$ is the first observation for the backward filter. Thus we first update (omitting the initial propagation) to obtain $\hat{\mathbf{y}}(t_N)$, and then propagate to obtain $\hat{\mathbf{y}}(t_{N-1}^+)$, etc.

Let us now apply the backward filter to the example of Section 3, namely

$$\dot{y}(t) = ay(t) \tag{8a}$$

$$y(t_{N+1}) = y_{N+1} \ . \tag{8b}$$

The system model is

$$\dot{Y}(t) = aY(t)$$

$$Y(t_{N+1}) = \hat{y}_{N+1}$$

where $\mathcal{E}\hat{y}_{N+1} = y_{N+1}$ and $\text{Var } \hat{y}_{N+1} = Q_{N+1}$. The measurement model is

$$Z_k = x(t_k) + v_k \ , \ k = 1, 2, \cdots, N$$

where $\mathcal{E}v_k = 0$ and $\mathcal{E}v_k^2 = R$ (independent of k).

Then with the Green's function $W(t,s) = e^{a(t-s)}$ and the notation $E = e^{a\Delta t}$, an application of the backward Kalman filter equations yields

$$\hat{y}(t_{N+1}) = \hat{y}_{N+1}$$

$$\hat{y}(t_N^+) = E^{-1}\hat{y}_{N+1}$$

$$\hat{y}(t_{N+1-j}^+) = E^{-1}\hat{y}(t_{N+2-j}) \ , \ j=2,3, \cdots ,N$$

$$\hat{y}(t_{N+1-j}) = Q(t_{N+1-j})[Q^{-1}(t_{N+1-j}^+)\hat{y}(t_{N+1-j}^+)$$
$$+ R^{-1}Z_{N+1-j}] \ , \ j=1,2, \cdots ,N$$

$$Q(t_{N+1}) = Q_{N+1}$$

$$Q(t_N^+) = E^{-2}Q_{N+1}$$

$$Q(t_{N+1-j}^+) = E^{-2}Q(t_{N+2-j}) \ , \ j=2,3, \cdots ,N$$

$$Q^{-1}(t_{N+1-j}) = Q^{-1}(t_{N+1-j}^+) + R^{-1}, \ j=1,2, \cdots ,N \ .$$

Inductively, or by identifying with the forward filter, we obtain

$$\hat{y}(t_n) = Q(t_n)E^{-n}\left[E^{N+1}Q_{N+1}^{-1}\hat{y}_{N+1} + R^{-1}\sum_{\alpha=n}^{N} E^{\alpha}Z_{\alpha} \right] \ , \ n=1,2, \cdots ,N$$

and

$$Q^{-1}(t_n) = E^{-2n}\left[E^{2(N+1)}Q_{N+1}^{-1} + R^{-1}\sum_{\alpha=n}^{N} E^{2\alpha} \right] \ , \ n=1,2, \cdots ,N \ .$$

Now suppose we have no information regarding the initial condition $\hat{y}_{N+1}$. Then $Q_{N+1}^{-1}=0$ and the value of $\hat{y}_{N+1}$ is immaterial. In this case

$$\hat{y}(t_n) = \frac{E^n\sum_{\alpha=n}^{N} E^{\alpha}Z_{\alpha}}{\sum_{\alpha=n}^{N} E^{2\alpha}} \ , \ n=1,2, \cdots ,N \tag{9}$$

and

$$Q(t_n) = \frac{R}{E^{-2n}\sum_{\alpha=n}^{N} E^{2\alpha}} \ , \ n=1,2, \cdots ,N \ . \tag{10}$$

If, furthermore, $a=0$, then

$$\hat{y}(t_n) = \frac{1}{N+1-n}\sum_{\alpha=n}^{N} Z_{\alpha} \ , \ n=1,2, \cdots ,N \tag{11}$$

and

$$Q(t_n) = \frac{1}{N+1-n} R \ , \ n=1,2, \cdots ,N \ . \tag{12}$$

5. THE SMOOTHED SOLUTION

One way in which the backward filter may be used in conjunction with the forward filter is as follows. Suppose $\hat{\mathbf{x}}(t_k^-)$ is an estimate of the state vector $\mathbf{x}$ at time $t = t_k$ obtained from the forward filter and $\hat{\mathbf{y}}(t_k^+)$ is an estimate of $\mathbf{x}$ at time $t = t_k$ obtained from the backward filter. Note that $\hat{\mathbf{x}}(t_k^-)$ and $\hat{\mathbf{y}}(t_k^+)$ are independent and neither uses the observation $\mathbf{Z}_k$. Now the corresponding updated estimates are

$$\hat{\mathbf{x}}(t_k) = \mathrm{BL}[\hat{\mathbf{x}}(t_k^-), \mathbf{Z}_k | \mathbf{x}(t_k)]$$

and

$$\hat{\mathbf{y}}(t_k) = \mathrm{BL}[\hat{\mathbf{y}}(t_k^+), \mathbf{Z}_k | \mathbf{x}(t_k)]$$

respectively, both of which, of course, use $\mathbf{Z}_k$.

Since $\hat{\mathbf{x}}(t_k^-)$ and $\hat{\mathbf{y}}(t_k)$ are independent we may consider

$$\mathrm{BL}[\hat{\mathbf{x}}(t_k^-), \hat{\mathbf{y}}(t_k) | \mathbf{x}(t_k)] \tag{1a}$$

—an improved estimate of $\mathbf{x}(t_k)$. But $\hat{\mathbf{x}}(t_k)$ and $\hat{\mathbf{y}}(t_k^+)$ also are independent, and therefore

$$\mathrm{BL}[\hat{\mathbf{x}}(t_k), \hat{\mathbf{y}}(t_k^+) | \mathbf{x}(t_k)] \tag{1b}$$

is also an improved estimate of $\mathbf{x}(t_k)$. Furthermore, both (1a) and (1b) use the entire data set $\mathbf{Z}_1, \cdots, \mathbf{Z}_N$ (with no redundancies). But from Property (ii) of the BL operator we know that (1a) and (1b) are equal with probability one. Thus we may write (1) as

$$\begin{aligned}\hat{\mathbf{s}}(t_k) &= M(t_k)[P^{-1}(t_k^-)\hat{\mathbf{x}}(t_k^-) + Q^{-1}(t_k)\hat{\mathbf{y}}(t_k)] \\ &= M(t_k)[P^{-1}(t_k)\hat{\mathbf{x}}(t_k) + Q^{-1}(t_k^+)\hat{\mathbf{y}}(t_k^+)]\end{aligned} \tag{2}$$

where

$$\begin{aligned}M^{-1}(t_k) &= P^{-1}(t_k^-) + Q^{-1}(t_k) \\ &= P^{-1}(t_k) + Q^{-1}(t_k^+)\end{aligned} \tag{3}$$

is its inverse covariance matrix. The operation of combining two unbiased estimates of $\mathbf{x}(t_k)$ is called *smoothing*.

We also see that $\hat{\mathbf{x}}(t_k^-)$, $\hat{\mathbf{y}}(t_k^+)$, and $\mathbf{Z}_k$ all are independent and use the entire data set $\mathbf{Z}_1, \cdots, \mathbf{Z}_N$ (with no redundancies). Thus, again from Property (ii) of the BL operator

$$\mathrm{BL}[\hat{\mathbf{x}}(t_k^-), \hat{\mathbf{y}}(t_k^+), \mathbf{Z}_k | \mathbf{x}(t_k)] \tag{4}$$

is also identical with (1), and

$$M^{-1}(t_k) = P^{-1}(t_k^-) + Q^{-1}(t_k^+) + H_k'R_k^{-1}H_k \tag{5}$$

[see (3)]. We conclude, therefore, that it is immaterial whether we update the backward filter and then smooth, or update the forward filter and then smooth, or smooth and then update the smoothed estimate.

Suppose now we smooth *both* updated estimates,

$$\hat{\boldsymbol{\xi}}(t_k) = \mathrm{BL}[\hat{\mathbf{x}}(t_k), \hat{\mathbf{y}}(t_k) | \mathbf{x}(t_k)] \ . \tag{6}$$

The estimates $\hat{\mathbf{x}}(t_k)$ and $\hat{\mathbf{y}}(t_k)$ are now correlated since both utilize $\mathbf{Z}_k$. The cross-covariance matrix of $\hat{\mathbf{x}}(t_k)$ and $\hat{\mathbf{y}}(t_k)$ is

$$\mathrm{Cov}(\hat{\mathbf{x}}(t_k), \hat{\mathbf{y}}(t_k)) = P(t_k)(H_k'R_k^{-1}H_k)Q(t_k) \tag{7}$$

and hence [see (25) of Section 1] the covariance matrix $\Xi(t_k)$ of $\hat{\boldsymbol{\xi}}(t_k)$ is given by

$$\Xi(t_k) = (\mathbf{H}'\mathbf{R}^{-1}\mathbf{H})^{-1}$$

where

$$\mathbf{H}' = [I \quad I]$$

and

$$\mathbf{R} = \begin{bmatrix} P(t_k) & P(t_k)(H_k'R_k^{-1}H_k)Q(t_k) \\ Q(t_k)(H_k'R_k^{-1}H_k)P(t_k) & Q(t_k) \end{bmatrix} \tag{8}$$

are partitioned matricies. Actually $\hat{\boldsymbol{\xi}}(t_k)$ is a worse estimate of $\mathbf{x}(t_k)$ than is $\hat{s}(t_k)$—as we shall see in the following example.

Now let us consider the scalar example treated in the two previous sections. From (2)

$$\hat{s}(t_n) = M(t_n)[E^{-2}P^{-1}(t_{n-1})E\hat{x}(t_{n-1}) + Q^{-1}(t_n)y(t_n)] \tag{9}$$

$$= M(t_n)E^{-n}\left[P_0^{-1}\hat{x}_0 + E^{N+1}Q_{N+1}^{-1}\hat{y}_{N+1} + R^{-1}\sum_{\alpha=1}^{N} E^{\alpha}Z_{\alpha} \right]$$

and from (3), the inverse variance of $\hat{s}(t_n)$ is given by

$$M^{-1}(t_n) = E^{-2}P^{-1}(t_{n-1}) + Q^{-1}(t_n)$$

$$= E^{-2n}\left[P_0^{-1} + E^{2(N+1)}Q_{N+1}^{-1} + R^{-1}\sum_{\alpha=1}^{N} E^{2\alpha} \right] \ . \tag{10}$$

Under the assumption that $P_0^{-1} = 0 = Q_{N+1}^{-1}$, the above formulas reduce to

$$\hat{s}(t_n) = \frac{E^n \sum\limits_{\alpha=1}^{N} E^\alpha Z_\alpha}{\sum\limits_{\alpha=1}^{N} E^{2\alpha}} \tag{11}$$

and

$$M(t_n) = \frac{R}{E^{-2n} \sum\limits_{\alpha=1}^{N} E^{2\alpha}} \, , \tag{12}$$

while the additional assumption $a=0$ implies that

$$\hat{s}(t_n) = \frac{1}{N} \sum_{\alpha=1}^{N} Z_\alpha$$

and

$$M(t_n) = \frac{1}{N} R \ .$$

For this concrete example let us also calculate $\hat{\xi}(t_n)$[see (6)] and its variance $\Xi(t_n)$. From (7)

$$\mathrm{Cov} \ (\hat{x}(t_n), \hat{y}(t_n)) = P(t_n) R^{-1} Q(t_n)$$

and from (8)

$$\mathbf{R} = \begin{bmatrix} P(t_n) & R^{-1}P(t_n)Q(t_n) \\ R^{-1}P(t_n)Q(t_n) & Q(t_n) \end{bmatrix} \ .$$

Thus from (25) and (26) of Section 1,

$$\Xi^{-1}(t_n) = \frac{P(t_n) + Q(t_n) - 2R^{-1}P(t_n)Q(t_n)}{P(t_n)Q(t_n)[1 - R^{-2}P(t_n)Q(t_n)]} \tag{13}$$

while

$$\hat{\xi}(t_n) = \frac{[1 - R^{-1}P(t_n)]Q(t_n)\hat{x}(t_n) + [1 - R^{-1}Q(t_n)]P(t_n)\hat{y}(t_n)}{P(t_n) + Q(t_n) - 2R^{-1}P(t_n)Q(t_n)} \ . \tag{14}$$

If we make the assumptions $P_0^{-1} = 0 = Q_{N+1}^{-1}$ and $a=0$, then

$$\Xi(t_n) = \frac{R}{N-1} \left[1 - \frac{1}{n(N+1-n)} \right] \tag{15}$$

and

$$\hat{\xi}(t_n) = \left(\frac{n-1}{N-1}\right)\hat{x}(t_n) + \left(\frac{N-n}{N-1}\right)\hat{y}(t_n) \ . \tag{16}$$

Thus if $n = N$,

$$\hat{\xi}(t_n) = \hat{x}(t_n)$$

$$\Xi(t_n) = P(t_n)$$

and if $n = 1$,

$$\hat{\xi}(t_n) = \hat{y}(t_n)$$

$$\Xi(t_n) = Q(t_n) \ ,$$

while if $1 \neq n \neq N$,

$$\Xi(t_n) = \operatorname{Var} \hat{\xi}(t_n) > \operatorname{Var} \hat{x}(t_n) = P(t_n)$$

$$\Xi(t_n) = \operatorname{Var} \hat{\xi}(t_n) > \operatorname{Var} \hat{y}(t_n) = Q(t_n) \ .$$

6. RELATION BETWEEN LEAST SQUARES AND KALMAN FILTERING

We assert that every least squares problem may be interpreted as a special case of the forward Kalman filter. That is, least squares theory is a subset of Kalman filtering. Before showing the equivalence, however, let us briefly recall the least squares formulation.

The classical least squares problem is generally presented in the following form. Let

$$z_j = \mathbf{h}_j'\mathbf{x} + v_j \ , \ 1 \leq j \leq n \tag{1}$$

be n independent scalar measurements on the p-dimensional parameter vector $\mathbf{x}$. We assume that $\mathbf{h}_j$ is a vector of known constants, that $\mathscr{E}v_j = 0$ and that $\mathscr{E}v_j^2 = \sigma^2 > 0$ (independent of j). With this information we wish to determine an unbiased estimate $\mathbf{x}^*$ of $\mathbf{x}$ which is best in the sense that the sum of the squares of the residuals, namely

$$\sum_{j=1}^{n}(z_j - \mathbf{h}_j'\mathbf{x})^2$$

is minimized. We call $\mathbf{x}^*$ the *least squares estimate* (LSE) of $\mathbf{x}$.

Now (1) may be written as the single vector equation

$$\mathbf{z} = H\mathbf{x} + \boldsymbol{\nu} \tag{2}$$

where $\mathbf{z}' = \{z_1, \cdots, z_n\}$, $\boldsymbol{\nu}' = \{\nu_1, \cdots, \nu_n\}$, and $H' = [\mathbf{h}_1 \cdots \mathbf{h}_n]$ is assumed to have rank p. The random vector $\boldsymbol{\nu}$ has mean zero and positive definite covariance matrix $\sigma^2 I$ where I is the identity matrix. Then the LSE $\mathbf{x}^*$ of $\mathbf{x}$ is

$$\mathbf{x}^* = (H'H)^{-1}H'\mathbf{z} \tag{3}$$

with covariance matrix

$$P^* = \sigma^2(H'H)^{-1} . \tag{4}$$

If we generalize (2) so that $\boldsymbol{\nu}$ has arbitrary positive definite covariance matrix Σ, then the least squares solution is still given by (3), but its covariance matrix is now

$$(H'H)^{-1}H\Sigma H(H'H)^{-1} .$$

However, the *Markov estimate* (ME) $\hat{\mathbf{x}}$ of $\mathbf{x}$ is

$$\hat{\mathbf{x}} = (H'\Sigma^{-1}H)^{-1}H'\Sigma^{-1}\mathbf{z} \tag{5}$$

and its covariance matrix is

$$P = (H'\Sigma^{-1}H)^{-1} . \tag{6}$$

[See (1) and (2) of Section 1.] The Gauss-Markov theorem states that if $\Sigma = \sigma^2 I$, then the LSE and ME are equal.

Let us make a trivial generalization of (1) or (2). Suppose

$$\mathbf{z}_k = H_k\mathbf{x} + \boldsymbol{\nu}_k , \ 1 \leq k \leq m \tag{7}$$

are m independent vector measurements on the parameter vector $\mathbf{x}$. We assume that

$$\mathbf{H}' = [H_1' \cdots H_m'] \tag{8}$$

has rank p, and that the noise vector $\boldsymbol{\nu}_k$ has mean zero and positive definite covariance matrix R_k. Then the BLUE of $\mathbf{x}$ is

$$\hat{\mathbf{x}} = P\sum_{k=1}^{m} H_k'R_k^{-1}\mathbf{z}_k \tag{9}$$

where

$$P = \left[\sum_{k=1}^{m} H_k'R_k^{-1}H_k \right]^{-1} \tag{10}$$

is the covariance matrix of $\hat{\mathbf{x}}$. [See (10) and (9) of Section 1.] Of course we may write (9) and (10) more compactly in the ME form, viz.:

$$\hat{\mathbf{x}} = P\mathbf{H}'\mathbf{R}^{-1}\mathbf{Z} \tag{11}$$

and

$$P = (\mathbf{H}'\mathbf{R}^{-1}\mathbf{H})^{-1} \tag{12}$$

where $\mathbf{Z}' = \{\mathbf{z}_1', \cdots, \mathbf{z}_m'\}$, $\mathbf{H}$ is given by (8), and $\mathbf{R}$ is a block diagonal matrix of the R_k matrices. Comparing (11) and (12) with (5) and (6) it is seen that no generalization is to be gained by allowing $\mathbf{R}$ to be an arbitrary positive definite covariance matrix—we only get the ME with a larger number of dependent observations. [See (26) and (25) of Section 1.]

We turn now to the Kalman formulation. Let $t_0, t_1, \cdots, t_m$ be a strictly increasing sequence of time points and consider the Kalman system model

$$\dot{\mathbf{X}} = \mathbf{0}$$
$$\mathbf{X}(t_0) = \hat{\mathbf{x}}_0 \tag{13}$$

where the initial condition $\hat{\mathbf{x}}_0$ has covariance matrix P_0. Let

$$\mathbf{Z}_k = H_k\mathbf{x}(t_k) + \boldsymbol{v}_k \ , \ 1 \leq k \leq m \ , \tag{14}$$

be the measurement model where $\mathbf{H}$ [see(8)] has rank p and the mean zero noise vector $\boldsymbol{v}_k$ has positive definite covariance matrix R_k. That is, we are choosing the measurement model to be identical with the least squares model of (7). If we apply the forward filter equations of Section 3 to this model, then the BLUE $\hat{\mathbf{x}}(t_m)$ of $\mathbf{x}(t_m)$ is given by

$$\hat{\mathbf{x}}(t_m) = P(t_m)\left[P_0^{-1}\hat{\mathbf{x}}_0 + \sum_{k=1}^{m} H_k'R_k^{-1}\mathbf{Z}_k \right] \tag{15}$$

where

$$P(t_m) = \left[P_0^{-1} + \sum_{k=1}^{m} H_k'R_k^{-1}H_k \right]^{-1} \tag{16}$$

is the covariance matrix of $\hat{\mathbf{x}}(t_m)$. But these equations are identical with (9) and (10) if we let $P_0^{-1} = 0$.

Thus we see that every least squares problem is equivalent to a Kalman filter. The measurement model of the Kalman filter is identical with the observation model of the least squares problem. The system model of the Kalman filter is

$$\dot{\mathbf{X}} = 0$$

$$\mathbf{X}(t_0) = \hat{\mathbf{x}}_0$$

where the inverse covariance matrix of $\hat{\mathbf{x}}_0$ is zero.

7. THE NONHOMOGENEOUS EQUATION

In our derivation of the forward and backward Kalman filter equations in Sections 3 and 4 we assumed that the deterministic equation satisfied by the state vector (that is, the differential equation governing the physical phenomenon) was a *homogeneous* equation. Thus, for example, our theory, as presented above, is not immediately applicable to even so simple a differential equation as

$$\dot{\mathbf{x}}(t) = \mathbf{g}(t)$$

where $\mathbf{g}(t)$ is a given function. In this section we shall show how the nonhomogeneous problem may be reduced to the homogeneous cases previously studied.

Suppose, then, that $F(t)$ and $G(t)$ are square matrices and that $\mathbf{g}(t)$ is a vector function, all continuous on the closed finite interval T. Then if t_0 is any point in T and $\mathbf{x}_0$ any constant vector, we may find the unique solution to the differential system.

$$\dot{\mathbf{x}}(t) = F(t)\mathbf{x}(t) + G(t)\mathbf{g}(t) \tag{1a}$$

$$\mathbf{x}(t_0) = \mathbf{x}_0 \tag{1b}$$

—say by use of the one-sided Green's function matrix $W(t,s)$ associated with $F(t)$. Now if we introduce the new variable $\boldsymbol{\xi}(t)$ defined by

$$\boldsymbol{\xi}(t) = \mathbf{x}(t) - \boldsymbol{\gamma}(t) \tag{2}$$

where

$$\boldsymbol{\gamma}(t) = \int_{t_0}^{t} W(t,s)G(s)\mathbf{g}(s)\,ds \;, \tag{3}$$

then $\boldsymbol{\xi}(t)$ satisfies the homogeneous equation

$$\dot{\boldsymbol{\xi}}(t) = F(t)\boldsymbol{\xi}(t)$$

with the same initial condition as (1b), viz.:

$$\boldsymbol{\xi}(t_0) = \mathbf{x}_0 \;.$$

In the Kalman theory the system model corresponding to (1) is

$$\dot{\mathbf{X}}(t) = F(t)\mathbf{X}(t) + G(t)\mathbf{g}(t) + \mathbf{V}(t) \tag{4}$$

$$\mathbf{X}(t_0) = \hat{\mathbf{x}}_0$$

where $\mathbf{V}(t)$ has mean zero and $\mathscr{E}\hat{\mathbf{x}}_0 = \mathbf{x}_0$. The deterministic forcing function $\mathbf{g}(t)$ is called a *control input*. As usual, we shall take our measurement model to be of the form

$$\mathbf{z}_k = H_k \mathbf{x}(t_k) + \boldsymbol{\nu}_k \ , \ k = 1, 2, \cdots \tag{5}$$

where $\boldsymbol{\nu}_k$ has mean zero. If we now make the transformations

$$\boldsymbol{\Xi}(t) = \mathbf{X}(t) - \boldsymbol{\gamma}(t) \tag{6}$$

$$\boldsymbol{\zeta}_k = \mathbf{z}_k - H_k \boldsymbol{\gamma}(t_k) \tag{7}$$

where $\boldsymbol{\gamma}$ has been defined by (2), then the system model (4) becomes

$$\dot{\boldsymbol{\Xi}}(t) = F(t)\boldsymbol{\Xi}(t) + \mathbf{V}(t) \tag{8}$$

$$\boldsymbol{\Xi}(t_0) = \hat{\mathbf{x}}_0$$

and the measurement model (5) becomes

$$\boldsymbol{\zeta}_k = H_k \boldsymbol{\xi}(t_k) + \boldsymbol{\nu}_k \ , \ k = 1, 2, \cdots \ . \tag{9}$$

Equations (8) and (9) are now in the standard form treated in Section 3 and subsequently.

We consider an example. Let

$$\dot{x}(t) = c$$

$$x(t_0) = x_0$$

be the deterministic system satisfied by a scalar state vector $x(t)$. Let

$$\dot{X}(t) = c$$

$$X(t_0) = \hat{x}_0$$

be the corresponding system model where $\mathscr{E}\hat{x}_0 = x_0$ and $\mathrm{Var}\ \hat{x}_0 = P_0$. (We are assuming the system noise is zero.) For the measurement model we choose

$$z_k = x(t_k) + \nu_k \ , \ k = 1, 2, \cdots, N \ , \tag{10}$$

where $\mathscr{E}\nu_k = 0$ and $\mathscr{E}\nu_k^2 = \sigma^2$ (independent of k).

Now the Green's function $W(t,s)$ for this problem is one, so that [see (3)]

$$\boldsymbol{\gamma}(t) = c(t - t_0) \ .$$

Hence if we make the definitions

$$\xi(t) = x(t) - c(t - t_0)$$
$$\Xi(t) = X(t) - c(t - t_0) \tag{11}$$
$$\zeta_k = z_k - c(t_k - t_0)$$

[see (2), (6), (7)] then in terms of these new variables

$$\dot{\xi}(t) = 0$$
$$\xi(t_0) = x_0$$

and our problem reduces to:

System model

$$\dot{\Xi}(t) = 0$$
$$\Xi(t_0) = \hat{x}_0$$

Measurement model

$$\zeta_k = \xi(t_k) + v_k , \ k = 1, 2, \cdots, N ,$$

where $\mathcal{E}\hat{x}_0 = x_0$, $\mathrm{Var}\ \hat{x}_0 = P_0$, $\mathcal{E}v_k = 0$, $\mathrm{Var}\ v_k = \sigma^2$.

But this is now the same problem treated in Section 3 corresponding to the special case $a = 0$. Hence [with $P_0^{-1} = 0$, see (24) of Section 3]

$$\hat{\xi}(t_n) = \frac{1}{n} \sum_{\alpha=1}^{n} \zeta_\alpha , \ n = 1, 2, \cdots, N , \tag{12}$$

and [see (25) of Section 3] the variance $\Pi(t_n)$ of $\hat{\xi}(t_n)$ is

$$\Pi(t_n) = \frac{1}{n} \sigma^2 , \ n = 1, 2, \cdots, N .$$

To translate (12) back to the original variables, we use (11) to write

$$\hat{x}(t_n) - c(t_n - t_0) = \frac{1}{n} \sum_{\alpha=1}^{n} [z_\alpha - c(t_\alpha - t_0)] , \tag{13}$$

and if we let

$$\Delta t = t_{k+1} - t_k ,$$

(independent of k) then (13) reduces to

$$\hat{x}(t_n) = \tfrac{1}{2}(n - 1)c\Delta t + \frac{1}{n} \sum_{\alpha=1}^{n} z_\alpha . \tag{14}$$

Its variance $P(t_n)$ is

$$P(t_n) = \frac{1}{n} \sigma^2 \ . \tag{15}$$

Now let us construct the backward filter for this problem. The deterministic differential system is

$$\dot{y}(t) = c$$

$$y(t_{N+1}) = y_{N+1}$$

with system model

$$\dot{Y}(t) = c$$

$$Y(t_{N+1}) = \hat{y}_{N+1}$$

where $\mathcal{E}\hat{y}_{N+1} = x(t_{N+1})$ and $\text{Var } \hat{y}_{N+1} = Q_{N+1}$. The measurement model is

$$z_k = y(t_k) + v_k \ , \ k = 1, 2, \cdots, N \ .$$

For this case $\gamma(t)$ is $-c(t_{N+1} - t)$, and hence we define:

$$\eta(t) = y(t) + c(t_{N+1} - t)$$

$$H(t) = Y(t) + c(t_{N+1} - t) \tag{16}$$

$$\zeta_k' = z_k + c(t_{N+1} - t_k) \ .$$

Then in terms of these new variables

$$\dot{\eta}(t) = 0$$

$$\eta(t_{N+1}) = y_{N+1}$$

and our problem reduces to:

System model

$$\dot{H}(t) = 0$$

$$H(t_{N+1}) = \hat{y}_{N+1}$$

Measurement model

$$\zeta_k' = \eta(t_k) + v_k \ , \ k = 1, 2, \cdots, N \ ,$$

where $\mathcal{E}\hat{y}_{N+1} = y_{N+1} = x(t_{N+1})$, $\text{Var } \hat{y}_{N+1} = Q_{N+1}$, $\mathcal{E}v_k = 0$, $\text{Var } v_k = \sigma^2$.

Thus from (11) of Section 4 ($a = 0$, $Q_{N+1}^{-1} = 0$)

$$\hat{\eta}(t_n) = \frac{1}{N+1-n} \sum_{\alpha=n}^{N} \zeta'_\alpha \, , \, n = 1,2, \cdots ,N \, , \tag{17}$$

with variance

$$Y(t_n) = \frac{1}{N+1-n} \sigma^2 \, , \, n = 1,2, \cdots ,N \, , \tag{18}$$

[see (12) of Section 4]. From (16), we may write (17) in terms of the original variables as

$$\hat{y}(t_n) + c(t_{N+1} - t_n) = \frac{1}{N+1-n} \sum_{\alpha=n}^{N} [z_\alpha + c(t_{N+1} - t_\alpha)]$$

or

$$\hat{y}(t_n) = -\tfrac{1}{2}(N-n)c\Delta t + \frac{1}{N+1-n} \sum_{\alpha=n}^{N} z_\alpha \, . \tag{19}$$

The variance $Q(t_n)$ of $\hat{y}(t_n)$ is

$$Q(t_n) = \frac{1}{N+1-n} \sigma^2 \, . \tag{20}$$

We also may determine the smoothed estimate. Since from (14)

$$\hat{x}(t_n^-) = \tfrac{1}{2}nc\Delta t + \frac{1}{n-1} \sum_{\alpha=1}^{n-1} z_\alpha$$

the smoothed estimate is

$$\hat{s}(t_n) = \mathrm{BL}[\hat{x}(t_n^-), \hat{y}(t_n) | x(t_n)]$$

$$= [n - \tfrac{1}{2}(N+1)]c\Delta t + \frac{1}{N} \sum_{\alpha=1}^{N} z_\alpha$$

with variance

$$M(t_n) = \frac{1}{N}\sigma^2 \, .$$

Note that we cannot adapt the final smoothing equations of Section 5 for this example since ξ, ζ_k and η, ζ'_k are defined differently.

8. THE EXTENDED KALMAN FILTER

In practical applications of estimation theory one often encounters models in which nonlinearities are present. The theory of nonlinear estimation has been developed for special cases; but its practical application usually requires that the underlying probability density functions explicitly be included in the estimation algorithm [2, 26]. This latter constraint constitutes a requirement that often is impossible to satisfy, or is computationally expensive to implement. The difficulties in the nonlinear case lead one to look for linear models which provide a suitable approximation to the nonlinear equations. It is an accepted practice to approximate the nonlinear models by suitably chosen linear ones. The linearization is performed with respect to a computed state vector estimate. The Kalman equations for propagating and updating the state vector and its covariance matrix, although derived under assumptions of linearity, are typically applied as an approximation to the nonlinear case [20, 32]. One usually refers to the resulting equation as an *extended Kalman filter* (EKF).

The rationale behind this approach is heuristic. It is motivated by the fact that the noise processes in the System and Measurement models are usually unknown; and the numbers used for their respective covariance matrices (in numerical implementations) are determined empirically. Linearizing the nonlinear Kalman models, however, has the effect of introducing unmodeled errors. If one is especially interested in precise parameter estimation (minimum system noise), and not merely data smoothing (for example, pointing and tracking), the effects of nonlinearities in the implementation algorithm should be analyzed, and, if possible, any spurious behavior corrected [38].

Let $\mathbf{x}(t)$ be a state vector which satisfies the deterministic ordinary differential system

$$\dot{\mathbf{x}}(t) = \mathbf{f}(\mathbf{x}(t),t) \tag{1a}$$

$$\mathbf{x}(t_0) = \mathbf{x}_0 \ . \tag{1b}$$

It will be assumed that $\mathbf{f}(\cdot,\cdot)$ is a smooth vector function with domain $J \times T$ where J is a p-dimensional cube (p being the dimension of $\mathbf{x}$) and T is a closed finite interval. In the initial condition (1b) we, of course, assume that $t_0 \in T$.

In the Kalman theory we shall consider the nonlinear system model

$$\dot{\mathbf{X}}(t) = \mathbf{f}(\mathbf{X}(t),t) + \mathbf{V}(t) \tag{2a}$$

$$\mathbf{X}(t_0) = \hat{\mathbf{x}}_0 \tag{2b}$$

where $\mathbf{V}(t)$ has mean zero and cross-covariance matrix $S(t,s) = \Omega(t)\delta(t-s)$ where $\Omega(t)$ is a nonnegative definite spectral density matrix. Thus $\mathbf{V}(t)$ is (nonstationary) white noise. The random initial condition (2b) will be assumed to have mean $\mathbf{x}_0$ and positive definite covariance matrix P_0.

The measurement model we shall use is

$$\mathbf{Z}_k = \mathbf{h}_k(\mathbf{x}(t_k)) + \boldsymbol{v}_k \ , \ k = 1,2, \cdots$$

where $\mathbf{x}$ is the p-dimensional state vector, $\mathbf{Z}_k$ an N-dimensional observation vector, and $\boldsymbol{v}_k$ a mean zero random vector with $n \times n$ positive definite covariance matrix R_k. We also shall assume that, in an appropriate domain, $\mathbf{h}_k$ is a smooth function. Generally the existence of second derivatives is sufficient.

It is further assumed that $\mathbf{V}(t)$, $\hat{\mathbf{x}}_0$, and $\boldsymbol{v}_k$, $k = 1,2, \cdots$ are independent.

Under these assumptions, the equations for propagating the state vector and its covariance matrix are as follows [20].

State estimate propagation

$$\dot{\hat{\mathbf{x}}}(t) = \mathbf{f}(\hat{\mathbf{x}}(t),t) \ , \ t_0 \leqq t < t_1$$

$$\hat{\mathbf{x}}(t_0) = \hat{\mathbf{x}}_0$$

and

$$\dot{\hat{\mathbf{x}}}(t) = \mathbf{f}(\hat{\mathbf{x}}(t),t)$$

has initial condition $\hat{\mathbf{x}}(t_k)$ in $[t_k, t_{k+1})$, $k = 1,2, \cdots$.

Error covariance propagation

$$\dot{P}(t) = F(\hat{\mathbf{x}}(t),t)P(t) + P(t)F'(\hat{\mathbf{x}}(t),t) + \Omega(t) \ , \ t_0 \leqq t < t_1$$

$$P(t_0) = P_0$$

and

$$\dot{P}(t) = F(\hat{\mathbf{x}}(t),t)P(t) + P(t)F'(\hat{\mathbf{x}}(t),t) + \Omega(t)$$

has initial condition $P(t_k)$ in $[t_k, t_{k+1})$, $k = 1,2, \cdots$.

In the above equations

$$F(\hat{\mathbf{x}}(t),t) = \left. \frac{\partial \mathbf{f}(\mathbf{x}(t),t)}{\partial \mathbf{x}'(t)} \right|_{\mathbf{x}(t) = \hat{\mathbf{x}}(t)} .$$

State estimate update

$$\hat{\mathbf{x}}(t_k) = \hat{\mathbf{x}}(t_k^-) + K_k[\mathbf{Z}_k - \mathbf{h}_k(\hat{\mathbf{x}}(t_k^-))] \ , \ k = 1,2, \cdots$$

Error covariance update

$$P(t_k) = [I - K_k H_k(\hat{\mathbf{x}}(t_k^-))]P(t_k^-) \ , \ k = 1,2, \cdots$$

Gain matrix

$$K_k = P(t_k^-)H_k'(\hat{\mathbf{x}}(t_k^-))[H_k(\hat{\mathbf{x}}(t_k^-))P(t_k^-)H_k'(\hat{\mathbf{x}}(t_k^-)) + R_k]^{-1} \ .$$

In the above equations

$$H_k(\hat{\mathbf{x}}(t_k^-)) = \frac{\partial \mathbf{h}_k(\mathbf{x}(t))}{\partial \mathbf{x}'(t)}\bigg|_{\mathbf{x}(t) = \hat{\mathbf{x}}(t_k^-)} \ .$$

The Kalman filter treated earlier in this chapter, as well as the EKF just discussed, are examples of a *continuous* system model and a *discrete* measurement model. This is probably the most commonly encountered type. However, one could also consider a discrete system model (that is, a difference equation), say

$$\mathbf{X}_k = F_{k-1}\mathbf{X}_{k-1} + \mathbf{V}_k$$

and a continuous measurement model, say

$$\mathbf{Z}(t) = H(t)\mathbf{X}(t) + \boldsymbol{v}(t) \ .$$

One also could use system and measurement models which were both continuous, as well as their EKF versions. These and many other more exotic combinations have been treated in the literature [20].

Chapter *II*

A Trajectory System Model

0. INTRODUCTION

In Chapter I we derived the basic Kalman formulas, and indicated how to extend these results to cope with nonlinear system and measurement models. Various simple examples were given to illustrate the general theory. In this chapter we would like to derive the system model for a sophisticated dynamical system. The important nontrivial problem we have in mind is the one alluded to in the Preface and in Section 3 of Chapter I—that is, the estimation of the trajectory of a ballistic reentry vehicle. (In Chapter III we shall derive the associated measurement models.)

The dimension of the state vector (as well as the complexity of the system model) depends on how realistically we choose our physical model. We shall derive, in a rigorous manner, the differential equation for the motion of a rigid body together with the associated aerodynamics. The effects of gravity, drag, and lift will be considered. Higher order effects such as a nonspherical gravity potential, vehicle angle of attack, and aerodynamic asymmetries will be included. If only gravity is present in the system model, then the system's state vector will be six-dimensional for a point mass, or twelve-dimensional for a point mass and rigid body. These cases correspond, respectively, to a three degrees of freedom and a six degrees of freedom model. The inclusion of aerodynamics will cause the state vector to be augmented according to the number of aerodynamic effects present. Naturally, more sophisticated models will lead to higher dimensional state vectors and more complicated differential equations for the system model. Even the most optimistic of readers would hardly expect the resulting differential equation to be linear. Hence the results on non-linear models must be invoked.

The analysis will proceed in three stages. In Section 1 the basic equations of motion for a point mass will be derived together with the associated aerodynamic effects; and in Section 2 the dynamics of a rigid body will be studied. Aerodynamic considerations will be treated further in greater detail in Section 3. The results of this chapter then will be summarized in Section 4 in the context of a Kalman system model.

We digress to make a remark on notation. If $\mathbf{Q}$ is a nonzero vector, then we shall write $\hat{Q}$ to indicate a unit vector in the direction of $\mathbf{Q}$. Thus

$$\hat{Q} = \frac{1}{|\mathbf{Q}|}\,\mathbf{Q}$$

where $|\mathbf{Q}|$ is the Euclidean norm (length) of $\mathbf{Q}$. This, of course, is at variance with the notation of Chapter I. No confusion will arise, however, since in deriving the differential equations of the system model we shall not be concerned with random variables and statistical estimation.

1. EQUATIONS OF MOTION OF A POINT MASS

We consider the motion of a point mass P, the vehicle's center of mass, moving along a trajectory C with respect to an earth-centered inertial (ECI) coordinate frame of reference $(\mathbf{X}, \mathbf{Y}, \mathbf{Z})$, see Fig. 1. The vectors $\mathbf{X}, \mathbf{Y}, \mathbf{Z}$ form a right-handed coordinate system with $\mathbf{X}, \mathbf{Y}$ in the equatorial plane and $\mathbf{Z}$ pointing through true North. (The vector $\mathbf{Z}$ lies along the earth's spin axis.) The point O represents the center of the earth. The position vector $\mathbf{r}(t)$ from O to P is

$$\mathbf{r}(t) = x(t)\hat{X} + y(t)\hat{Y} + z(t)\hat{Z}$$

and the velocity vector $\mathbf{V}$ is

$$\mathbf{V}(t) \equiv \dot{\mathbf{r}}(t) = \dot{x}(t)\hat{X} + \dot{y}(t)\hat{Y} + \dot{z}(t)\hat{Z}\ . \tag{1}$$

Since the $\hat{X}, \hat{Y}, \hat{Z}$ vectors form an ECI coordinate system, they are not functions of time, t. Hence

$$\dot{\hat{X}} = \dot{\hat{Y}} = \dot{\hat{Z}} = \mathbf{0}\ .$$

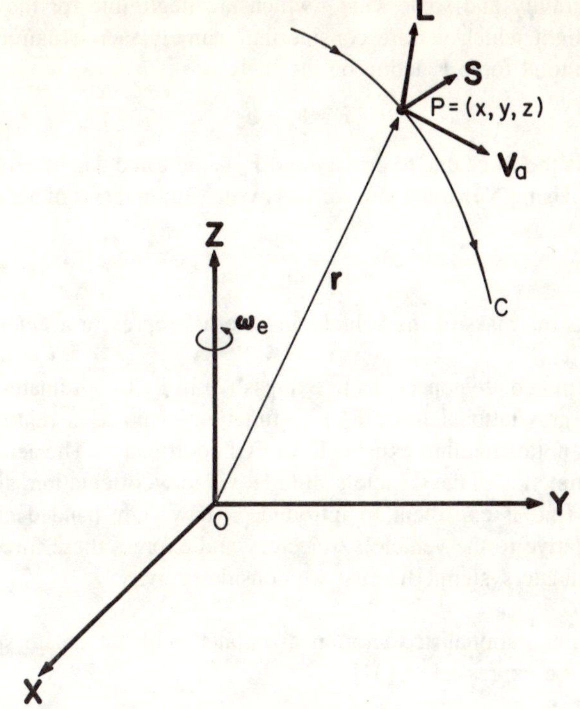

Figure 1
Trajectory of Vehicle

For convenience we introduce the triple, u, v, w where

$$u(t) \equiv \dot{x}(t)$$

$$v(t) \equiv \dot{y}(t)$$

$$w(t) \equiv \dot{z}(t) \ .$$

Thus, see (1)

$$\mathbf{V}(t) = u(t)\hat{X} + v(t)\hat{Y} + w(t)\hat{Z} \ .$$

Note that $\hat{V}$ is a unit vector tangent to the trajectory C at the point P, while $\mathbf{W} = \mathbf{r} \times \mathbf{V}$ is the binormal to C at P, and $\mathbf{N} = \mathbf{W} \times \mathbf{V}$ is the normal to C at P. The vectors $\hat{V}, \hat{N}, \hat{W}$ form a right-handed coordinate system. The (principal) normal lies in the osculating plane to C at P, and the binormal is perpendicular to the osculating plane.

The external forces acting on the vehicle are assumed to be due to the earth's gravity and aerodynamics. We neglect higher order effects such as lunar gravity and solar wind—which are negligible for the phase of ballistic flight which we are considering, namely, aerodynamic reentry. Hence the total force $\mathbf{F}$ acting on the body is

$$\mathbf{F} = \mathbf{F}_g + \dot{\mathbf{F}}_a \tag{2a}$$

where $\mathbf{F}_g$ is the force due to gravity and $\mathbf{F}_a$ is the force due to aerodynamic pressures. Using Newton's law we may write (2a) in terms of accelerations as

$$m\mathbf{a} = m\mathbf{a}_g + m\mathbf{a}_a \tag{2b}$$

where m is the mass of the vehicle and the $\mathbf{a}$'s represent accelerations.

Our immediate concern is to express (2) in ECI coordinates $\hat{X}$, $\hat{Y}$, $\hat{Z}$. Since the gravitational force $\mathbf{F}_g$ is a function of position relative to the earth, it is not difficult to express $\mathbf{F}_g$ in ECI coordinates. The aerodynamic force is a function of the vehicle's altitude, velocity, orientation, and shape. We shall find it expedient to introduce a new right-handed coordinate system relative to the vehicle's trajectory and express these forces in this new coordinate system. But first we consider gravity.

The gravitational acceleration associated with an oblate spheroidal earth may be expressed as [31]

$$\mathbf{a}_g = g_R \hat{r} + g^* \hat{Z} \ .$$

If g_X, g_Y, g_Z are the ECI components of $\mathbf{a}_g$, then

$$g_X = g_R \frac{x}{|\mathbf{r}|}$$

$$g_Y = g_R \frac{y}{|\mathbf{r}|}$$

$$g_Z = g_R \frac{z}{|\mathbf{r}|} + g^*$$

and

$$\mathbf{a}_g = g_X \hat{X} + g_Y \hat{Y} + g_Z \hat{Z} \ . \tag{3}$$

The values of g_R and g^* are intimately related to the real world. If we adopt the usual expansion for the gravity potential [12], we have, to second order,

$$g_R = -\frac{GM}{|\mathbf{r}|^2}\left[1 + J\left(\frac{r_e}{|\mathbf{r}|}\right)^2 (1 - 5\sin^2\lambda) \right]$$

$$g^* = -\frac{2(GM)}{|\mathbf{r}|^2} J \left(\frac{r_e}{|\mathbf{r}|}\right)^2 \sin\lambda$$

where, based on the 1972 World Geodetic Survey,

$$GM = 3.986032 \times 10^5 \text{ km}^3/\text{sec}$$

$$= \text{earth-mass gravitational constant}$$

$$J = 1.624 \times 10^{-3}$$

$$= \text{gravitational harmonic constant}$$

$$r_e = 6,378.135 \text{ km}$$

$$= \text{earth equatorial radius}.$$

The angle λ is the geocentric latitude of the point P (see Chapter III),

$$\lambda = \arcsin \frac{z}{|\mathbf{r}|} \ .$$

We now consider the areodynamic force $\mathbf{F}_a$ in (2a). As the vehicle moves through the atmosphere there is some net force which is the result of the integrated pressure forces acting on the vehicle's body. This aerodynamic force depends on the atmospheric density and flow relative to the vehicle's body, and upon the size and shape of the body as well. It is customary to express the aerodynamic force in the form [12]

$$\mathbf{F}_a = qA\mathbf{C}_a \tag{4}$$

where q is the magnitude of the dynamic pressure acting on the body, A is the vehicle's effective drag area, and $\mathbf{C}_a$ is a vector whose components consist of dimensionless parameters. The magnitude of the dynamic pressure is given by

$$q = \tfrac{1}{2}\rho|\mathbf{V}_a|^2$$

where ρ is the atmospheric mass density at the vehicle, and $\mathbf{V}_a$ is the velocity of the vehicle relative to the surrounding air. If we assume that the atmosphere rotates with the earth (without slippage or wind) then

$$\mathbf{V}_a = \dot{\mathbf{r}} - \boldsymbol{\omega}_e \times \mathbf{r} \equiv \mathbf{V} - \boldsymbol{\omega}_e \times \mathbf{r} \ . \tag{5}$$

In the above equation $\boldsymbol{\omega}_e = \omega_e \hat{\mathbf{Z}}$ where ω_e ($= 7.292115147 \times 10^{-5}$ rad/sec) is the angular rotational rate of the earth about its axis $\mathbf{Z}$.

To treat the aerodynamic force in component form we introduce the wind axis triad $\mathbf{V}_a$, $\mathbf{S}$, $\mathbf{L}$ where $\mathbf{V}_a$ is given by (5),

$$\mathbf{S} = \mathbf{r} \times \mathbf{V}_a$$

$$\mathbf{L} = \mathbf{V}_a \times \mathbf{S}.$$

The wind axis triad provides the basis vectors for a moving coordinate frame, centered at the vehicle (the point P), and defined by the unit vectors $\hat{V}_a$, $\hat{S}$, $\hat{L}$—a right-handed coordinate system (see Fig. 1). Equation (4) now may be written in component form. Let

$$\mathbf{F}_a = -F_D \hat{V}_a + F_S \hat{S} + F_L \hat{L} \ . \tag{6}$$

If we write

$$\mathbf{C}_a = -C_D \hat{V}_a + C_S \hat{S} + C_L \hat{L} \ , \tag{7}$$

then using (4) we have

$$F_D = C_D A q$$

$$F_S = C_S A q \tag{8}$$

$$F_L = C_L A q \ .$$

The definitions for the components of $\mathbf{F}_a$ given above correspond to the conventional ones for drag, side, and lift aerodynamic forces. The vector sum of the side and lift forces is the total lift force $\mathbf{F}_\ell$,

$$\mathbf{F}_\ell = F_S \hat{S} + F_L \hat{L} \ .$$

The aerodynamic parameters C_D, C_S, C_L correspond, respectively, to the drag, side, and lift forces acting on the vehicle. They are, in turn, functions of the atmospheric density and flow relative to the vehicle's body, as well as the size and shape of the body. Their derivation from an aerodynamic design point of view does not concern us here. We are concerned only with their determination based on measurements on the vehicle's dynamics. In Section 3, after considerations of the six degrees of freedom rigid body dynamics, we shall be able to relate the trajectory oriented aerodynamic parameters C_D, C_S, C_L to other parameters that are body oriented.

As stated before, our immediate goal is to write (2) in terms of $\hat{X}$, $\hat{Y}$, $\hat{Z}$. Towards this end we must deduce the coordinate transformation between $\hat{X}$, $\hat{Y}$, $\hat{Z}$ and $\hat{V}_a$, $\hat{S}$, $\hat{L}$. Now

$$\mathbf{r} = x\hat{X} + y\hat{Y} + z\hat{Z}$$

$$\dot{\mathbf{r}} = u\hat{X} + v\hat{Y} + w\hat{Z}$$

and from (5)

$$\mathbf{V}_a = (u + \omega_e y)\hat{X} + (v - \omega_e x)\hat{Y} + w\hat{Z} \ . \tag{9}$$

It is convenient to introduce the suggestive notation $(Q)_A$, $(Q)_B$, $(Q)_C$ to indicate the components of a vector $\mathbf{Q}$ in an $\mathbf{A}$, $\mathbf{B}$, $\mathbf{C}$ coordinate frame, viz.:

$$\mathbf{Q} = (Q)_A\hat{A} + (Q)_B\hat{B} + (Q)_C\hat{C} \ .$$

Thus we may write (9) as

$$\mathbf{V}_a = (V_a)_X\hat{X} + (V_a)_Y\hat{Y} + (V_a)_Z\hat{Z} \ ,$$

Also

$$\mathbf{S} = \mathbf{r} \times \mathbf{V}_a$$

$$= (yw - zv + \omega_e xz)\hat{X} + (zu - xw + \omega_e yz)\hat{Y} + [xv - yu - \omega_e(x^2 + y^2)]\hat{Z}$$

$$\equiv (S)_X\hat{X} + (S)_Y\hat{Y} + (S)_Z\hat{Z}$$

and

$$\mathbf{L} = \mathbf{V}_a \times \mathbf{S}$$

$$= [(V_a)_Y(S)_Z - (V_a)_Z(S)_Y]\hat{X}$$

$$+ [(V_a)_Z(S)_X - (V_a)_X(S)_Z]\hat{Y}$$

$$+ [(V_a)_X(S)_Y - (V_a)_Y(S)_X]\hat{Z}$$

$$\equiv (L)_X\hat{X} + (L)_Y\hat{Y} + (L)_Z\hat{Z} \ .$$

Thus

$$\begin{bmatrix} \hat{V}_a \\ \hat{S} \\ \hat{L} \end{bmatrix} = \mathbf{D} \begin{bmatrix} \hat{X} \\ \hat{Y} \\ \hat{Z} \end{bmatrix} \tag{10}$$

where

$$\mathbf{D} = \begin{bmatrix} \dfrac{(V_a)_X}{|\mathbf{V}_a|} & \dfrac{(V_a)_Y}{|\mathbf{V}_a|} & \dfrac{(V_a)_Z}{|\mathbf{V}_a|} \\[2ex] \dfrac{(S)_X}{|\mathbf{S}|} & \dfrac{(S)_Y}{|\mathbf{S}|} & \dfrac{(S)_Z}{|\mathbf{S}|} \\[2ex] \dfrac{(L)_X}{|\mathbf{L}|} & \dfrac{(L)_Y}{|\mathbf{L}|} & \dfrac{(L)_Z}{|\mathbf{L}|} \end{bmatrix}$$

is an orthogonal matrix.

Hence, from (2), using (3) and (6), we have the representation

$$
m \begin{bmatrix} \ddot{x} \\ \ddot{y} \\ \ddot{z} \end{bmatrix} = m \begin{bmatrix} g_X \\ g_Y \\ g_Z \end{bmatrix} + \mathbf{D}' \begin{bmatrix} -F_D \\ F_S \\ F_L \end{bmatrix} \tag{11}
$$

where

$$
\mathbf{F} = m\mathbf{a} = m\ddot{\mathbf{r}} = m(\ddot{x}\hat{X} + \ddot{y}\hat{Y} + \ddot{z}\hat{Z}).
$$

In deducing (10) from (9) the following lemma may clarify the reduction.

Lemma. Let $(\hat{\Xi}, \hat{H}, \hat{Z})$ and $(\hat{A}, \hat{B}, \hat{C})$ be orthogonal right-handed coordinate systems with the same origin. Let Q be a point with coordinates (ξ, η, ζ) in the $(\hat{\Xi}, \hat{H}, \hat{Z})$ frame and coordinates (α, β, γ) in the $(\hat{A}, \hat{B}, \hat{C})$ frame. Let Θ be a nonsingular transformation from $(\hat{\Xi}, \hat{H}, \hat{Z})$ to $(\hat{A}, \hat{B}, \hat{C})$,

$$
\begin{bmatrix} \hat{\Xi} \\ \hat{H} \\ \hat{Z} \end{bmatrix} = \Theta \begin{bmatrix} \hat{A} \\ \hat{B} \\ \hat{C} \end{bmatrix} .
$$

Then Θ is orthogonal and

$$
\begin{bmatrix} \xi \\ \eta \\ \zeta \end{bmatrix} = \Theta \begin{bmatrix} \alpha \\ \beta \\ \gamma \end{bmatrix} .
$$

Our next task is to reduce (11) to a first order differential equation. First, using (7) we may write (11) as

$$
\begin{bmatrix} \ddot{x} \\ \ddot{y} \\ \ddot{z} \end{bmatrix} = \begin{bmatrix} g_X \\ g_Y \\ g_Z \end{bmatrix} + \frac{Aq}{m} \mathbf{D}' \begin{bmatrix} -C_D \\ C_S \\ C_L \end{bmatrix} . \tag{12}
$$

Let

$$
\{x, y, z, u, v, w\}
$$

be the state vector. Then since $\{\ddot{x}, \ddot{y}, \ddot{z}\} = \{\dot{u}, \dot{v}, \dot{w}\}$ we may write (12) as

$$
\begin{bmatrix} \dot{x} \\ \dot{y} \\ \dot{z} \\ \dot{u} \\ \dot{v} \\ \dot{w} \end{bmatrix} = \begin{bmatrix} u \\ v \\ w \\ \begin{bmatrix} g_X \\ g_Y \\ g_Z \end{bmatrix} + \dfrac{Aq}{m} \mathbf{D}' \begin{bmatrix} -C_D \\ C_S \\ C_L \end{bmatrix} \end{bmatrix} . \tag{13}
$$

If we assume a spherical earth, then $J = 0$ and

$$g_R = -\frac{GM}{|\mathbf{r}|^2}$$

$$g^* = 0 \ .$$

Thus

$$g_X = -\frac{GM}{r^3}\, x$$

$$g_Y = -\frac{GM}{r^3}\, y$$

$$g_Z = -\frac{GM}{r^3}\, z$$

where

$$r = |\mathbf{r}| = \sqrt{x^2 + y^2 + z^2} \ .$$

If we assume drag, but no side or lift forces, then in (13),

$$C_S = 0$$

$$C_L = 0$$

and

$$\mathbf{D}' \begin{bmatrix} -C_D \\ C_S \\ C_L \end{bmatrix} = -\frac{C_D}{|\mathbf{V}_a|} \begin{bmatrix} u + \omega_e y \\ v - \omega_e x \\ w \end{bmatrix} \ .$$

Thus since $q = \frac{1}{2}\rho |\mathbf{V}_a|^2$, eq. (13) reduces to

$$\begin{bmatrix} \dot{x} \\ \dot{y} \\ \dot{z} \\ \dot{u} \\ \dot{v} \\ \dot{w} \end{bmatrix} = \begin{bmatrix} u \\ v \\ w \\ -\dfrac{GMx}{r^3} - \dfrac{\rho A}{2m} C_D |\mathbf{V}_a|(u + \omega_e y) \\ -\dfrac{GMy}{r^3} - \dfrac{\rho A}{2m} C_D |\mathbf{V}_a|(v - \omega_e x) \\ -\dfrac{GMz}{r^3} - \dfrac{\rho A}{2m} C_D |\mathbf{V}_a|(w) \end{bmatrix} \ . \qquad (14)$$

2. EQUATIONS OF MOTION OF A RIGID BODY

We consider in this section the orientation and rotation of a rigid body. We recall from the methods of classical dynamics that any motion of a rigid body may be represented by the superposition of two elementary motions: a translation of the center of mass, and a rotation about an instantaneous axis through the center of mass [42]. Thus, by combining the results of this section with those of the previous section, we shall be able to describe completely the dynamics of a (rigid) reentry vehicle.

We find it convenient to describe the rotational dynamics in terms of two special coordinate frames: a body-centered inertial (BCI) coordinate frame of reference and a body-fixed rotating frame. Let vectors $\mathbf{X}_b, \mathbf{Y}_b, \mathbf{Z}_b$ lie parallel, respectively, to $\mathbf{X}, \mathbf{Y}, \mathbf{Z}$ of the ECI coordinate frame. The BCI coordinate frame is defined as a right-handed coordinate frame with vectors $\mathbf{X}_b, \mathbf{Y}_b, \mathbf{Z}_b$ and centered at P, the vehicle's center of mass. We note that the BCI coordinate frame is not truly inertial. It moves (translates) with respect to the ECI coordinate frame along C, the trajectory of P. Hence we shall restrict our use of the BCI coordinate frame to considerations of rotational motion only.

A body-fixed coordinate frame may be defined by considering the vehicle's inertial mass distribution. We shall assume that the vehicle's body is elongated from its base to nose and is also nearly symmetric with respect to an axis pointing along this major dimension (the vehicle's major axis). We also shall assume that the vehicle's mass distribution is similar to its body's shape. These assumptions imply that a stable rotational motion exists when the vehicle is spinning about its major axis. Also, if $\mathbf{I}$ is the moment of inertia tensor of the vehicle, then its eigenvalues, say I_R, I_P, I_Y, are distinct and the corresponding eigenvectors, say $\mathbf{R}_{OL}, \mathbf{P}_{IT}, \mathbf{Y}_{AW}$, are orthogonal. The shape of the vehicle's mass distribution implies that one eigenvector, say $\mathbf{R}_{OL}$, points toward the nose of the vehicle, while the remaining two, $\mathbf{P}_{IT}$ and $\mathbf{Y}_{AW}$ lie in a transverse plane [4].

Let the eigenvectors of $\mathbf{I}$ be ordered so that $(\mathbf{R}_{OL}, \mathbf{P}_{IT}, \mathbf{Y}_{AW})$ form a right-handed coordinate frame centered at P, see Fig. 2. As the notation suggests, the $\mathbf{R}_{OL}, \mathbf{P}_{IT}, \mathbf{Y}_{AW}$ axes define, respectively, the orientation of roll, pitch, and yaw rotations of the vehicle. Since the orientations of $\mathbf{R}_{OL}, \mathbf{P}_{IT}, \mathbf{Y}_{AW}$ (the eigenvectors of $\mathbf{I}$) are invariant with respect to the vehicle, they constitute a body-fixed coordinate frame and rotate with the vehicle at its angular velocity, $\boldsymbol{\omega}$. We call this coordinate frame the vehicle's principal axes (VPA).

An immediate problem is to define the relation between the BCI and VPI coordinate frames. This relation is provided via the classical Eulerian parametrization of an arbitrary finite rotation [21]. Let θ, ϕ, ψ be the Euler angles transforming the VPA coordinate frame to the BCI coordinate frame.

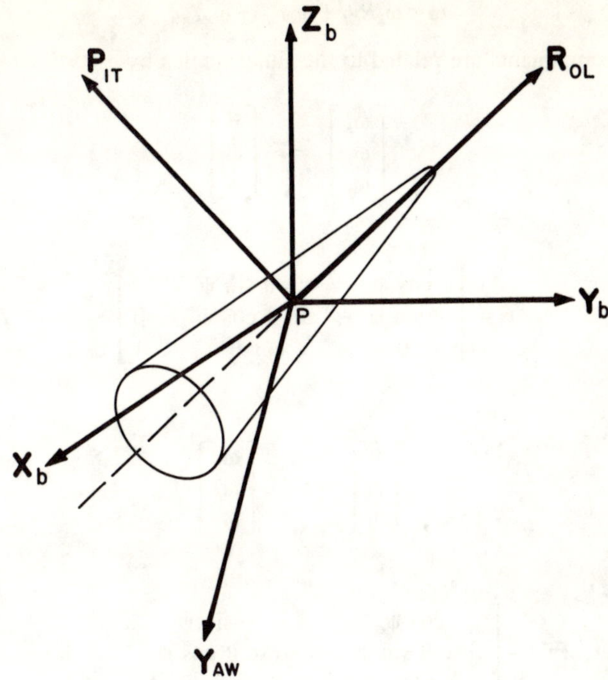

Figure 2
Vehicle Orientation

Then

$$\begin{bmatrix} \hat{R}_{OL} \\ \hat{P}_{IT} \\ \hat{Y}_{AW} \end{bmatrix} = \mathbf{E} \begin{bmatrix} \hat{X}_b \\ \hat{Y}_b \\ \hat{Z}_b \end{bmatrix} \tag{1}$$

where

$$\mathbf{E} = \begin{bmatrix} \cos\phi\cos\psi \\ -\cos\theta\sin\phi\sin\psi & \sin\phi\cos\psi \\ +\cos\theta\cos\phi\sin\psi & \sin\theta\sin\psi \\[2ex] -\cos\phi\sin\psi \\ -\cos\theta\sin\phi\cos\psi & -\sin\phi\sin\psi \\ +\cos\theta\cos\phi\cos\psi & \sin\theta\cos\psi \\[2ex] \sin\theta\sin\phi & -\sin\theta\cos\phi & \cos\theta \end{bmatrix} \tag{2}$$

is the Eulerian angle transformation. Now if the vehicle is rotating at an angular velocity $\boldsymbol{\omega}$, then the orientation of the VPA frame relative to the BCI coordinate frame, changes. Hence $\mathbf{E}$ is a function of time. Let the components of $\boldsymbol{\omega}$ with respect to the VPA coordinate frame be ω_R, ω_P, ω_Y,

$$\boldsymbol{\omega} = \omega_R \hat{R}_{OL} + \omega_P \hat{P}_{IT} + \omega_Y \hat{Y}_{AW} \ .$$

These components are related to the Euler angles by

$$\begin{bmatrix} \omega_R \\ \omega_P \\ \omega_Y \end{bmatrix} = \mathbf{F} \begin{bmatrix} \dot{\theta} \\ \dot{\phi} \\ \dot{\psi} \end{bmatrix}$$

where

$$\mathbf{F} = \begin{bmatrix} \cos\psi & \sin\theta\,\sin\psi & 0 \\ -\sin\psi & \sin\theta\,\cos\psi & 0 \\ 0 & \cos\theta & 1 \end{bmatrix} \ . \tag{3}$$

Thus

$$\begin{bmatrix} \dot{\theta} \\ \dot{\phi} \\ \dot{\psi} \end{bmatrix} = \mathbf{F}^{-1} \begin{bmatrix} \omega_R \\ \omega_P \\ \omega_Y \end{bmatrix} \tag{4}$$

where

$$\mathbf{F}^{-1} = \begin{bmatrix} \cos\psi & -\sin\psi & 0 \\ \csc\theta\,\sin\psi & \csc\theta\,\cos\psi & 0 \\ -\cot\theta\,\sin\psi & -\cot\theta\,\cos\psi & 1 \end{bmatrix} \ . \tag{5}$$

We observe that the transformation matrix $\mathbf{F}$ is singular for $\theta = 0$, although the VPA and BCI coordinate frames enjoy a well-defined (physical) relationship. For $\theta = 0$ we have

$$\mathbf{E}\Big|_{\theta=0} = \begin{bmatrix} \cos(\phi+\psi) & \sin(\phi+\psi) & 0 \\ -\sin(\phi+\psi) & \cos(\phi+\psi) & 0 \\ 0 & 0 & 1 \end{bmatrix} \ .$$

That is, ϕ and ψ cannot uniquely be determined when $\theta = 0$. This singularity is an artifact of the Eulerian parametrization which uses three independent rotation angles. Other methods exist whereby the singularity may be removed, for example, four parameter representations (quaternions) or direction cosines. For more on these methods see [15, 18] .

The principle of rotational dynamics states that [21]

$$\mathbf{G} = \frac{d\mathbf{H}}{dt} \tag{6}$$

where $\mathbf{H}$ is the angular momentum and $\mathbf{G}$ is the sum of the external torques . [Equation (6) is analogous to $\mathbf{F} = d\mathbf{M}/dt$ where $\mathbf{M} = m\mathbf{V}$ is the momentum ($m =$ mass, $\mathbf{V} =$ velocity) and $\mathbf{F}$ is force .] Both $\mathbf{H}$ and $\mathbf{G}$ are taken with respect to the center of mass, P, of the vehicle .   The angular momentum $\mathbf{H}$ is related to the angular velocity $\boldsymbol{\omega}$ by the vehicle's moment of inertia tensor $\mathbf{I}$,

$$\mathbf{H} = \mathbf{I}\boldsymbol{\omega} \ . \tag{7}$$

When $\mathbf{I}$ is referenced to the BCI coordinate frame, the components of its matrix ($\mathbf{I} \equiv I_{\mathrm{BCI}}$) vary as the vehicle rotates. However, if $\mathbf{I}$ is referenced to its eigenvectors, the VPA frame, then its matrix ($\mathbf{I} \equiv I_{\mathrm{VPA}}$) is diagonal and constant. The diagonal terms are its eigenvalues,

$$\mathbf{I} = I_{\mathrm{VPA}} = \begin{bmatrix} I_R & 0 & 0 \\ 0 & I_P & 0 \\ 0 & 0 & I_Y \end{bmatrix} \ .$$

The terms I_R, I_P, I_Y are the vehicle's moments of inertia about the $\mathbf{R}_{OL}$, $\mathbf{P}_{IT}$, $\mathbf{Y}_{AW}$ axes, respectively. Thus if (7) is expressed with respect to the VPA frame, we have

$$\begin{bmatrix} H_R \\ H_P \\ H_Y \end{bmatrix} = \begin{bmatrix} I_R & 0 & 0 \\ 0 & I_P & 0 \\ 0 & 0 & I_Y \end{bmatrix} \begin{bmatrix} \omega_R \\ \omega_P \\ \omega_Y \end{bmatrix} \tag{8}$$

where

$$\mathbf{H} = H_R \hat{R}_{OL} + H_P \hat{P}_{IT} + H_Y \hat{Y}_{AW} \ .$$

Now for any vector $\mathbf{Q}$, the derivative with respect to the nonrotating BCI coordinate frame and the derivative with respect to the rotating VPA frame are related by

$$\left(\frac{d\mathbf{Q}}{dt} \right)_{\mathrm{BCI}} = \left(\frac{d\mathbf{Q}}{dt} \right)_{\mathrm{VPA}} + (\boldsymbol{\omega} \times \mathbf{Q})_{\mathrm{VPA}} \tag{9}$$

where the subscripts BCI and VPA denote that the representations and differentiations are with respect to those coordinate frames. Also, as defined above, $\boldsymbol{\omega}$ is the rotation rate of the VPA frame relative to the BCI coordinate frame.

Returning to (6), we see that (9) implies

$$\mathbf{G} = \left(\frac{d\mathbf{H}}{dt} \right)_{\mathrm{BCI}} = \left(\frac{d\mathbf{H}}{dt} \right)_{\mathrm{VPA}} + (\boldsymbol{\omega} \times \mathbf{H})_{\mathrm{VPA}}$$

and from (7)

$$\mathbf{G} = \left(\frac{d\mathbf{I}\boldsymbol{\omega}}{dt} \right)_{\mathrm{BCI}} = \left(\frac{d\mathbf{I}\boldsymbol{\omega}}{dt} \right)_{\mathrm{VPA}} + (\boldsymbol{\omega} \times \mathbf{I}\boldsymbol{\omega})_{\mathrm{VPA}} \ .$$

But since the components of $\mathbf{I}$ are constant when expressed with respect to the VPA coordinate frame, we have

$$\mathbf{G} = I_{\text{VPA}}\left(\frac{d\boldsymbol{\omega}}{dt}\right)_{\text{VPA}} + (\boldsymbol{\omega} \times \mathbf{I}\boldsymbol{\omega})_{\text{VPA}}$$

or

$$G_R = I_R\dot{\omega}_R - (I_P - I_Y)\omega_P\omega_Y$$

$$G_P = I_P\dot{\omega}_P - (I_Y - I_R)\omega_Y\omega_R \qquad (10)$$

$$G_Y = I_Y\dot{\omega}_Y - (I_R - I_P)\omega_R\omega_P$$

where G_R, G_P, G_Y are the components of the torque $\mathbf{G}$ with respect to the roll, pitch, and yaw axes, respectively. From (10) we may write

$$\dot{\omega}_R = \left(\frac{I_P - I_Y}{I_R}\right)\omega_P\omega_Y + \frac{G_R}{I_R}$$

$$\dot{\omega}_P = \left(\frac{I_Y - I_R}{I_P}\right)\omega_Y\omega_R + \frac{G_P}{I_P} \qquad (11)$$

$$\dot{\omega}_Y = \left(\frac{I_R - I_P}{I_Y}\right)\omega_R\omega_P + \frac{G_Y}{I_Y} \, .$$

The torques G_R, G_P, G_Y are assumed to result from aerodynamic effects. Similar to the translational case discussed in the previous section, we assume that the torques are related to the aerodynamic parameters C_R, C_P, C_Y by equations of the form

$$G_R = C_R A q d$$

$$G_P = C_P A q d \qquad (12)$$

$$G_Y = C_Y A q d$$

where A is the vehicle's effective drag area, q is the magnitude of the dynamic pressure acting on the vehicle, and d is the vehicle's reference diameter. (In the case of a cone-shaped vehicle we would have $A = \pi d^2/4$.)

We now shall combine the results of the previous section with those derived here. The resulting first order differential equation will describe completely the general six degrees of freedom motion of the vehicle. Let

$$\boldsymbol{\xi}' = \{x, y, z, u, v, w, \theta, \phi, \psi, \omega_R, \omega_P, \omega_Y\}$$

be the twelve-dimensional state vector. Then

$$\dot{\boldsymbol{\xi}} \equiv \begin{bmatrix} \dot{\boldsymbol{\xi}}_1 \\ \dot{\theta} \\ \dot{\phi} \\ \dot{\psi} \\ \dot{\omega}_R \\ \dot{\omega}_P \\ \dot{\omega}_Y \end{bmatrix} = \begin{bmatrix} \mathbf{f}_1(\boldsymbol{\xi}) \\ \mathbf{F}^{-1}(\theta,\phi,\psi) \begin{bmatrix} \omega_R \\ \omega_P \\ \omega_Y \end{bmatrix} \\ \mathbf{g}_1(\boldsymbol{\xi}) \end{bmatrix} \tag{13}$$

where

$$\boldsymbol{\xi}_1' = \{x,\, y,\, z,\, u,\, v,\, w\}$$

and $\mathbf{f}_1$ is the right-hand side of (12) of Section 2 while $\mathbf{g}_1$ is the right-hand side of (11).

3. AERODYNAMIC CONSIDERATIONS

In this section we increase the number of aerodynamic parameters and express the aerodynamic force and torque parameters previously introduced, namely C_D, C_S, C_L, C_R, C_P, C_Y, in terms of more primitive aerodynamic effects. As before, we tacitly assume that the relevant aerodynamics may be lumped into a finite set of discrete parameters. Our point of view is that of trajectory reconstruction and parameter estimation.

To begin we need to relate the vehicle's orientation to the direction of its velocity relative to air. From (10) of Section 1 and (1) of Section 2 we may write this relationship as

$$\begin{bmatrix} \hat{V}_a \\ \hat{S} \\ \hat{L} \end{bmatrix} = \mathbf{DE}' \begin{bmatrix} \hat{R}_{OL} \\ \hat{P}_{IT} \\ \hat{Y}_{AW} \end{bmatrix} \tag{1}$$

where

$$\mathbf{D} \equiv \mathbf{D}(x,y,z,u,v,w)$$

$$\mathbf{E} \equiv \mathbf{E}(\theta,\phi,\psi)$$

are functions only of certain components of the state vector.

The transformation defined by (1) allows us to express the aerodynamic effects with respect to a principal plane of air flow, namely the $(\mathbf{R}_{OL}, \mathbf{V}_a)$ - plane. The angle between the $\mathbf{R}_{OL}$ and $\mathbf{V}_a$ vectors is called the total angle of attack, α. If we express $\mathbf{V}_a$ in terms of the VPA coordinate frame, say,

$$\mathbf{V}_a = V_1\hat{R}_{OL} + V_2\hat{P}_{IT} + V_3\hat{Y}_{AW} \; , \tag{2}$$

then

$$\cos\alpha = \frac{V_1}{|\mathbf{V}_a|} \; . \tag{3}$$

The components V_1, V_2, V_3 of $\mathbf{V}_a$ in the VPA system are given by the first row of $\mathbf{DE}'$ multiplied by the magnitude of $\mathbf{V}_a$.

Now define the vector $\mathbf{N}$ to be perpendicular to the $(\mathbf{R}_{OL},\mathbf{V}_a)$ - plane and perpendicular to $\mathbf{R}_{OL}$. We may describe $\mathbf{N}$ by the two equations

$$\mathbf{N} \cdot \mathbf{R}_{OL} = 0$$

and

$$(\mathbf{R}_{OL} \times \mathbf{V}_a) \times \mathbf{N} = \mathbf{0} \; . \tag{4}$$

Since $\mathbf{N}$ is perpendicular to $\mathbf{R}_{OL}$, it must be in the $(\mathbf{P}_{IT},\mathbf{Y}_{AW})$ - plane, and hence of the form

$$\hat{N} = D_P\hat{P}_{IT} + D_Y\hat{Y}_{AW} \; . \tag{5}$$

Equations (2), (4) and (5) imply that

$$D_P V_2 + D_Y V_3 = 0 \; .$$

Thus

$$D_P = \frac{V_3}{\sqrt{V_2^2 + V_3^2}}$$

$$D_Y = \frac{-V_2}{\sqrt{V_2^2 + V_3^2}} \; . \tag{6}$$

The aerodynamic force parameters C_D, C_S, C_L now may be expressed in terms of two primitive aerodynamic parameters: an axial drag parameter C_A, and a normal, side-force parameter, C_N. Since $\mathbf{V}_a$, $\mathbf{S}$, $\mathbf{L}$ and $\mathbf{R}_{OL}$, $\mathbf{P}_{IT}$, $\mathbf{Y}_{AW}$ both span the space [see (1)] we may write $\mathbf{C}_a$ [see (7) of Section 1] as

$$\mathbf{C}_a = C_A\hat{R}_{OL} + C_N\hat{N} \; .$$

Thus

$$\mathbf{C}_a \cdot \hat{V}_a = -C_D = C_A(\hat{R}_{OL} \cdot \hat{V}_a) + C_N(\hat{N} \cdot \hat{V}_a)$$

$$\mathbf{C}_a \cdot \hat{S} = C_S = C_A(\hat{R}_{OL} \cdot \hat{S}) + C_N(\hat{N} \cdot \hat{S}) \tag{7}$$

$$\mathbf{C}_a \cdot \hat{L} = C_L = C_A(\hat{R}_{OL} \cdot \hat{L}) + C_N(\hat{N} \cdot \hat{L}) \ .$$

By virtue of (1) and (6) the coefficients of C_A and C_N in the above expressions are known in terms of the components of the state vector.

We consider now the aerodynamic torque parameters for a stable aerodynamic reentry vehicle. In this case the primary torque is due to a misalignment between the roll axis $\mathbf{R}_{OL}$ and the velocity relative to air, $\mathbf{V}_a$. For a stable vehicle the center of pressure will be behind the center of mass, and hence the forces acting on the vehicle will tend to restore alignment of the $\mathbf{R}_{OL}$ and $\mathbf{V}_a$ vectors. This torque effect is proportional to α—for sufficiently small misalignments. A related, but higher-order effect, gives rise to a torque that is dependent upon the angle of attack rate, $\dot{\alpha}$. The other torque effect we shall include is proportional to the vehicle's angular velocity $\boldsymbol{\omega}$. We have, then,

$$C_R = C_{R\omega} \frac{\omega_R d}{2|\mathbf{V}_a|}$$

$$C_P = C_{P\omega} \frac{\omega_P d}{2|\mathbf{V}_a|} + C_{P\alpha}\alpha D_P + C_{P\dot{\alpha}} \frac{\dot{\alpha} D_P d}{2|\mathbf{V}_a|} \tag{8}$$

$$C_Y = C_{Y\omega} \frac{\omega_Y d}{2|\mathbf{V}_a|} + C_{Y\alpha}\alpha D_Y + C_{Y\dot{\alpha}} \frac{\dot{\alpha} D_Y d}{2|\mathbf{V}_a|} \ .$$

The C's on the right-hand side of (8) are the primitive aerodynamic torque parameters and are defined below. All the multiplicative factors except $\dot{\alpha}$ have been defined and discussed above and seen to depend only on the components of the state vector. The angle of attack rate $\dot{\alpha}$ is obtained by differentiating (3).

The aerodynamic coefficients are defined as:

$C_{R\omega}, C_{P\omega}, C_{Y\omega} = $ roll, pitch, yaw damping parameters due to angular velocity

$C_{P\alpha}, C_{Y\alpha} \quad = $ pitch and yaw torque parameters due to angle of attack

$C_{P\dot{\alpha}}, C_{Y\dot{\alpha}} \quad = $ pitch and yaw damping parameters due to angle of attack rate

Other higher-order effects such as vehicle aerodynamic trim or center of mass offset also may be included. We refer the reader to the literature for a more detailed discussion of aerodynamic effects [33, 47, 48].

4. THE KALMAN SYSTEM MODEL EQUATIONS

In Section 8 of Chapter I we wrote the system model for the nonlinear case in the form [see (2a) of that section],

$$\dot{\mathbf{X}}(t) = \mathbf{f}(\mathbf{X}(t),t) + \mathbf{V}(t) \ . \tag{1}$$

In applications pertaining to real-time radar tracking of a reentry vehicle the prime consideration is to successfully point the radar at the target. (See Chapter III for more on radar measurements.) In this case the rigid body dynamics may be dropped. The simplest situation is to let the state vector be

$$\mathbf{x}' = \{x,\ y,\ z,\ u,\ v,\ w,\ C_D\} \ .$$

The drag parameter C_D is related to the ballistic coefficient β by the equation

$$\beta = \frac{m}{AC_D} \ . \tag{2}$$

If, for simplicity in notation we let $\mathbf{X} = \mathbf{x}$, then our system model is

$$\dot{x} = u + \text{noise}$$

$$\dot{y} = v + \text{noise}$$

$$\dot{z} = w + \text{noise}$$

$$\dot{u} = g_x - \frac{\rho|\mathbf{V}_a|}{2\beta}\,(u + \omega_e y) + \text{noise} \tag{3}$$

$$\dot{v} = g_Y - \frac{\rho|\mathbf{V}_a|}{2\beta}\,(v - \omega_e x) + \text{noise}$$

$$\dot{w} = g_Z - \frac{\rho|\mathbf{V}_a|}{2\beta}\,(w) + \text{noise}$$

$$\dot{C}_D = 0 + \text{noise} \ .$$

The "noise" terms above are usually empirically obtained to allow specified filter performance despite real-time implementation short-cuts, such as approximations in gravity and atmospheric density models, as well as computer truncation errors.

In applications pertaining to post-mission trajectory reconstruction or vehicle accuracy assessment, all aerodynamic parameters might be included as further parameters to be estimated. In this case the state vector may be written as

$$\mathbf{x}' = \{x,\ y,\ z,\ u,\ v,\ w,\ \theta,\ \phi,\ \psi,\ \omega_R,\ \omega_P,\ \omega_Y,\ \mathbf{C}'\}$$

where

$$\mathbf{C}' = \{C_A,\ C_N,\ C_{R\omega},\ C_{P\omega},\ C_{Y\omega},\ C_{P\alpha},\ C_{Y\alpha},\ C_{P\dot{\alpha}},\ C_{Y\dot{\alpha}}\}$$

is a vector of unknown aerodynamic parameters. The system model is then

$$\dot{x} = u + \text{noise}$$

$$\dot{y} = v + \text{noise}$$

$$\dot{z} = w + \text{noise}$$

$$\begin{bmatrix} \dot{u} \\ \dot{v} \\ \dot{w} \end{bmatrix} = \begin{bmatrix} g_X \\ g_Y \\ g_Z \end{bmatrix} + \frac{Aq}{m}\,\mathbf{D}' \begin{bmatrix} -C_D \\ C_S \\ C_L \end{bmatrix} + \textbf{noise}$$

$$\begin{bmatrix} \dot{\theta} \\ \dot{\phi} \\ \dot{\psi} \end{bmatrix} = \mathbf{F}^{-1} \begin{bmatrix} \omega_R \\ \omega_P \\ \omega_Y \end{bmatrix} + \textbf{noise} \tag{4}$$

$$\dot{\omega}_R = \left(\frac{I_P - I_Y}{I_R} \right) \omega_P \omega_Y + \frac{C_R A q d}{I_R} + \text{noise}$$

$$\dot{\omega}_P = \left(\frac{I_Y - I_R}{I_P} \right) \omega_R \omega_Y + \frac{C_P A q d}{I_P} + \text{noise}$$

$$\dot{\omega}_Y = \left(\frac{I_R - I_P}{I_Y} \right) \omega_R \omega_P + \frac{C_Y A q d}{I_Y} + \text{noise}$$

$$\dot{\mathbf{C}} = \mathbf{0} + \textbf{noise}$$

where, again, for simplicity in notation we have written $\mathbf{X} = \mathbf{x}$.

In the case of post-mission processing, computer throughput usually is not a serious consideration. Thus extremely accurate gravity and atmospheric models may be employed, which allow system noise to be reduced and provide a more constrained fit to the data. Sometimes, however, system noise is needed even for exact equations like $\dot{x} = u$ to account for the finite precision of a computer [9].

The modeling of the drag parameter C_D in (3) as a constant, or the aerodynamic parameters $\mathbf{C}$ in (4) as a constant vector, is valid since these quantities tend to be slowly varying. The presence of system noise in their equations enables such a local solution to give rise to a global time varying solution—provided the data intervals are suitably chosen [6].

Chapter III

A Radar Measurement Model

0. INTRODUCTION

A radar is a measurement device which performs observations along a line-of-sight from itself to a target. The radar typically measures the time delay between a transmitted pulse and the corresponding echo, which results from the reflection of that transmitted pulse's energy off the target. The range, distance to the target, may be directly computed from this time delay. A radar's antenna is usually designed to be very sensitive to incoming energy along a particular axis, sometimes called the boresight, and to ignore incoming energy from other directions. This property is called the directivity of the antenna. It allows the pointing coordinates of the line-of-sight, such as azimuth and elevation, to be measured by finding the direction of maximum received energy relative to a reference platform. A radar is therefore capable of providing the position coordinates of a target.

The determination of a radar's site coordinates relative to another radar or other earth referenced platform, such as an earth-centered inertial (ECI) coordinate frame (see Chapter II) is more complex. For example, the irregularities of the earth's surface and mass distribution tend to preclude simplistic approaches from having more than approximate relevance to practical applications of orbit determination. Fortunately, however, rather simple models of the earth's surface, based on extensive measurements, particularly those recently derived via space-based platforms, can support practical applications of radar data to the determination of reentry vehicle trajectories. One such model is the earth reference ellipsoid. It is a close approximation to the geoid, the equipotential surface of gravity similar in shape to the mean sea level.

The definition of the earth reference ellipsoid will be presented in Section 1 and the conventions for describing positions relative to the earth's surface will be derived. The transformations needed to relate a reentry vehicle's position, expressed in earth-centered inertial coordinates, to radar observations on range, azimuth, and elevation will be derived in Section 2. In Section 3 the Kalman measurement model for radar observations is presented, and a discussion of correlated azimuth and elevation observations for monopulse radars is included. For completeness, some miscellaneous measurement models pertaining to on-board instrumentation are provided in Section 4.

1. EARTH REFERENCED COORDINATES

Let a radar be located at a point Q. We wish to express its coordinates relative to the earth's surface. With respect to a spherical earth model the coordinates of Q may be given in terms of the customary definitions of latitude and longitude. A third coordinate, the distance above or below the earth's surface, measured along a ray from the center of the earth to the radar, completes the unique specification of Q.

It is well established that the earth actually bulges near the equator and is flatter at the poles. An oblate spheroid, having appropriate dimensions, can therefore provide a more realistic model of the earth than a sphere. An equation for an oblate spheroid is

$$\frac{x^2 + y^2}{a^2} + \frac{z^2}{b^2} = 1$$

with $a > b > 0$.

We now provide the definition of the earth reference ellipsoid. Consider an ellipse centered at the earth's center with major axis lying in the equatorial plane and minor axis along the earth's rotation axis. The length of the semimajor axis a is defined as the earth's equatorial radius r_e. The length of the semiminor axis b is determined by the equation

$$b = a(1 - f)$$

where f is the flattening parameter of the ellipse. Since $b > a > 0$ we see that $0 < f < 1$. (The relation between f and the eccentricity

$$e = \frac{\sqrt{a^2 - b^2}}{a}$$

of the ellipse is therefore

$$e^2 = 2f - f^2 \quad . \;)$$

If we rotate this ellipse about its semiminor axis, the resultant ellipsoid of revolution will be the desired oblate spheroid. The cross-sections parallel to the equatorial plane are perfect circles (parallels of latitude). The earth's meridians are represented by ellipses having semimajor axes of length a, lying in the earth's equatorial plane, and semiminor axes of length b, lying along the earth's rotation axis. We complete the definition of the earth reference ellipsoid by providing values for the earth's equatorial radius and flattening parameter. Based on the 1972 World Geodetic Survey these are, respectively,

$$r_e = 6{,}378.135 \text{ km}$$

$$f = \frac{1}{298.26} \; .$$

The coordinates of longitude, latitude, and height above (or below) the earth's surface now need to be defined with respect to the earth reference ellipsoid. For longitude we retain the same definition as used with the spherical earth model.

Definition. *Longitude* is the angle lying in the equatorial plane, measured from the Greenwich meridian to the meridian of interest, with positive sense directed due East.

For latitude and height we must take into account the deviation of our earth model from that of a sphere. For latitude we have two useful definitions. We need consider only the ellipse in which the meridian containing the point of interest lies. The situation is illustrated in Fig. 1 (where the eccentricity of the ellipse has been exaggerated to clarify the definitions.)

Definition. The *geocentric latitude* λ is the acute angle measured from the semimajor axis to a line which connects the center of the ellipse (the earth's center) to the point of interest on the ellipse. The *geodetic latitude*

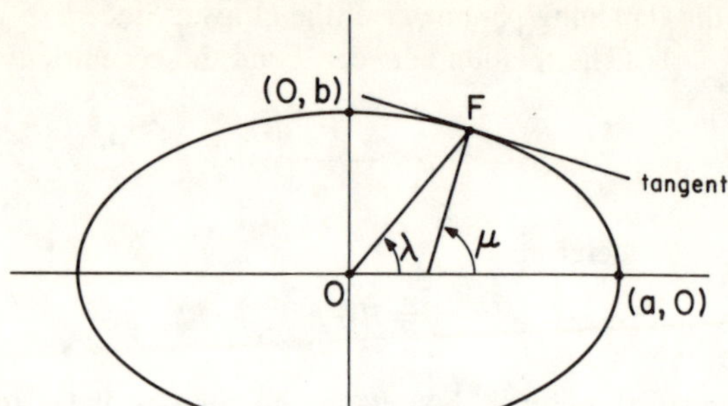

Figure 1
Geocentric and Geodetic Latitudes

μ is the acute angle measured from the semimajor axis to a line which connects the semimajor axis to the point of interest and is normal to the ellipse at that point.

We note that both definitions for latitude define angles which lie in planes that are perpendicular to the earth's equatorial plane and share the same semimajor axis.

We may relate these two definitions of latitude using the eccentricity e, or flattening parameter f, of the ellipse. If λ and μ represent, respectively, the geocentric and geodetic latitude of a point, then [12]

$$\mu = \arctan \left[\frac{1}{(1-f)^2} \tan \lambda \right] .$$

The *height* of an object above (or below) the earth reference ellipsoid is the distance from the ellipsoid to the point, measured along the vector that is normal to the ellipsoid and passes through the point. The point at which this vector intersects the ellipsoid is called the *footprint* of the object upon the ellipsoid. If the object is moving relative to the earth, then the curve on the ellipsoid generated by the footprint is called the *ground trace* or *ground track* of the object.

We now derive some useful relations for specifying points with respect to the earth reference ellipsoid. Suppose a radar is located at a point Q on the surface of the earth. Let its coordinates with respect to the earth reference ellipsoid be ℓ, λ, and h, namely, longitude, geocentric latitude, and

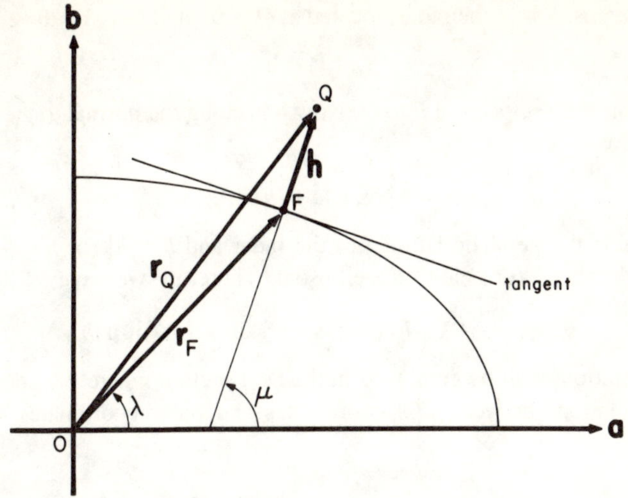

Figure 2
Earth Referenced Coordinates

height, respectively. Let the point F be the footprint of the radar on the earth reference ellipsoid. We choose for the origin of a coordinate system the center of the earth (the point O) and define $\mathbf{r}_Q$ as the vector from O to Q, and $\mathbf{r}_F$ as the vector from O to F, see Fig. 2. Now the local meridian containing F is congruent to the Greenwich meridian (the generating ellipse for the earth reference ellipsoid) but rotated about its minor axis by the angle ℓ (the longitude of the radar). Let $\mathbf{a}$ be a vector along the semimajor axis of the local meridian ellipse such that $\mathbf{a} \cdot \mathbf{r}_F$ is positive, and let $\mathbf{b}$ be a vector along the semiminor axis which pierces the North pole. The vectors $\mathbf{a}$ and $\mathbf{b}$ define a two-dimensional coordinate system containing the point F. Its origin is at O. The point Q also is contained in the plane defined by $\mathbf{a}$ and $\mathbf{b}$ since it lies along the normal at F to the ellipsoid of revolution generated by the Greenwich meridian.

Let the length of the vector $\mathbf{r}_F$ be r_F, the distance from O to F. Then

$$\mathbf{r}_F = (r_F \cos \lambda)\hat{a} + (r_F \sin \lambda)\hat{b} \ . \tag{1}$$

But from the properties of an ellipse it follows that

$$r_F = \frac{r_e \sqrt{1 - e^2}}{\sqrt{1 - e^2 \cos^2 \lambda}} \tag{2}$$

(where r_e is the earth's equatorial radius and e is the eccentricity of the ellipse).

The position of the radar, located at Q is related to $\mathbf{r}_F$ by the equation

$$\mathbf{r}_Q = \mathbf{r}_F + \mathbf{h} \tag{3}$$

where $\mathbf{h}$ is the vector from F to Q. Now $\mathbf{h}$ is along the normal to the ellipse at F. Thus

$$\mathbf{h} = (h\cos\mu)\hat{a} + (h\sin\mu)\hat{b} \tag{4}$$

where μ is the geodetic latitude of the radar and $h = |h|$ (or $h = -|h|$) is the height above (or below) the ellipsoid. Therefore we have

$$\mathbf{r}_Q = (r_F\cos\lambda + h\cos\mu)\hat{a} + (r_F\sin\lambda + h\sin\mu)\hat{b} \ . \tag{5}$$

These relations will be employed in the next section where the coordinate transformations between ECI coordinates and radar coordinates will be derived.

2. COORDINATE TRANSFORMATIONS

We consider a point P, the position of a reentry vehicle. Its coordinates with respect to the earth-centered inertial (ECI) coordinate frame of reference $(\mathbf{X}, \mathbf{Y}, \mathbf{Z})$ are (x, y, z), see Chapter II. The position vector $\mathbf{r}$ from the center of the earth (the point O) to P is

$$\mathbf{r} = x\hat{X} + y\hat{Y} + z\hat{Z} \ . \tag{1}$$

We recall that $\mathbf{X}$ and $\mathbf{Y}$ are in the equatorial plane and $\mathbf{Z}$ points through true North (along the earth's rotational axis). For definiteness we shall assume that $\mathbf{X}$ pierces the Greenwich meridian at midnight, see Fig. 3.

Measurements on range, azimuth, and elevation are made on P from a radar located at a point Q on the surface of the earth. The coordinates of the radar with respect to the earth reference ellipsoid are ℓ_0, λ, and h, the longitude, geocentric latitude, and height above the ellipsoid, respectively, of the point Q. The vector $\mathbf{r}_Q$ from O to Q has magnitude r_Q, the distance from the center of the earth to the radar.

Our objective is to express the range, azimuth, and elevation in ECI coordinates. To proceed most efficiently we first find the coordinates of $\mathbf{r}_Q$ relative to the ECI coordinate system. We then define a rectangular coordinate system $(\hat{X}_Q, \hat{Y}_Q, \hat{Z}_Q)$ relative to the radar and derive the matrix transformation from $(\hat{X}_Q, \hat{Y}_Q, \hat{Z}_Q)$ to the ECI basis $(\hat{X}, \hat{Y}, \hat{Z})$. Finally, if we write $\mathbf{R}$ in the $(\hat{X}_Q, \hat{Y}_Q, \hat{Z}_Q)$ coordinate frame, then range, azimuth, and elevation may be expressed simply in terms of the components of $\mathbf{R}$.

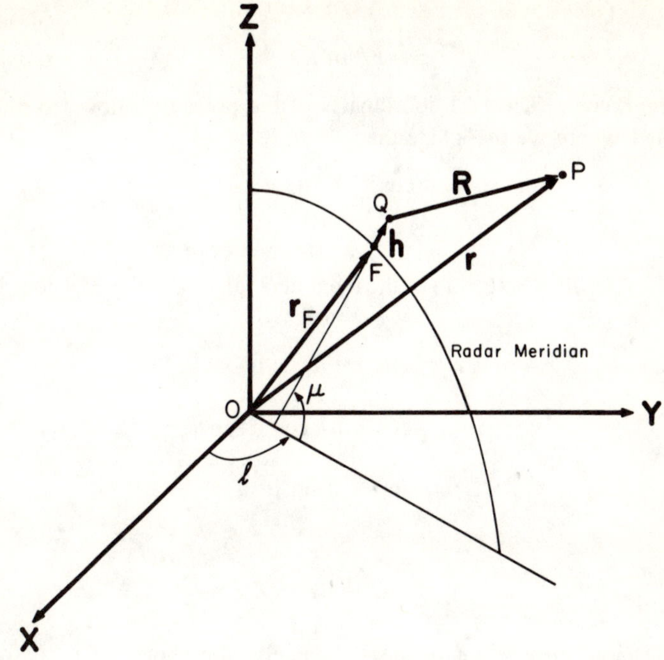

Figure 3
Radar and Vehicle Positions

Let the point F be the radar footprint on the earth reference ellipsoid. Let the vector $\mathbf{r}_F$ from O to F have coordinates (x_F, y_F, z_F) with respect to the ECI coordinate frame. Let the projection of $\mathbf{r}_F$ on the equatorial plane be

$$\mathbf{p}_F = x_F \hat{X} + y_F \hat{Y} \, . \tag{2a}$$

We define the ECI longitude of the radar as the angle ℓ,

$$\ell = \ell_0 + \omega_e t \, , \tag{3}$$

where t is the time after midnight and ω_e is the earth's rotational speed. Thus (2a) may be written as

$$\mathbf{p}_F = (p_F \cos \ell)\hat{X} + (p_F \sin \ell)\hat{Y} \tag{2b}$$

where $p_F = |\mathbf{p}_F|$.

Let $r_F = |\mathbf{r}_F|$ be the distance from the earth's center to F [see (2) of Section 1]. The ECI components of $\mathbf{r}_F$ are thus

$$x_F = r_F \cos \lambda \cos \ell$$

$$y_F = r_F \cos \lambda \sin \ell$$

$$z_F = r_F \sin \lambda$$

and $p_F = r_F \cos \lambda$. The ECI coordinates of the point Q follow from (5) of Section 1 where we recognize that

$$\hat{a} = (\cos \ell)\hat{X} + (\sin \ell)\hat{Y}$$

$$\hat{b} = \hat{Z}.$$

Thus the coordinates (ξ, η, ζ) of the radar at Q relative to the ECI coordinate frame are

$$\xi = (r_F \cos \lambda + h \cos \mu)\cos \ell$$

$$\eta = (r_F \cos \lambda + h \cos \mu)\sin \ell \tag{4}$$

$$\zeta = r_F \sin \lambda + h \sin \mu$$

and

$$\mathbf{r}_Q = \xi\hat{X} + \eta\hat{Y} + \zeta\hat{Z}. \tag{5}$$

We now turn to our next task, namely, the choice of a coordinate system relative to the radar. At the radar location Q choose a coordinate system whose vectors point East, North, and up, respectively, with respect to the earth reference ellipsoid (see Fig. 3). The origin of the coordinate system is at Q. We recall that the geodetic latitude defines a ray which is normal to the ellipsoid at the radar's footprint. Hence the ''up'' vector is given by

$$\hat{Z}_Q = (\cos \mu \cos \ell)\hat{X} + (\cos \mu \sin \ell)\hat{Y} + (\sin \mu)\hat{Z}. \tag{6}$$

The radar is assumed fixed to the earth and rotates with the earth about the **Z** axis. This enables us to derive the ''East'' and ''North'' vectors, respectively, as

$$\hat{X}_Q = \frac{\hat{Z} \times \hat{Z}_Q}{|\hat{Z} \times \hat{Z}_Q|}$$

$$= -(\sin \ell)\hat{X} + (\cos \ell)\hat{Y} \tag{7}$$

and

$$\hat{Y}_Q = \hat{Z}_Q \times \hat{X}_Q \tag{8}$$

$$= -(\sin \mu \cos \ell)\hat{X} - (\sin \mu \sin \ell)\hat{Y} + (\cos \mu)\hat{Z}.$$

We now may write the transformation matrix which relates the orientation of the ECI coordinate system's basis vectors to the $(\hat{X}_Q, \hat{Y}_Q, \hat{Z}_Q)$ basis vectors of the East-North-up (ENU) coordinate system. From (6), (7) and (8)

$$\begin{bmatrix} \hat{X}_Q \\ \hat{Y}_Q \\ \hat{Z}_Q \end{bmatrix} = \mathbf{A} \begin{bmatrix} \hat{X} \\ \hat{Y} \\ \hat{Z} \end{bmatrix} \tag{9}$$

where

$$\mathbf{A} = \begin{bmatrix} -\sin \ell & \cos \ell & 0 \\ -\sin \mu \cos \ell & -\sin \mu \sin \ell & \cos \mu \\ \cos \mu \cos \ell & \cos \mu \sin \ell & \sin \mu \end{bmatrix} \tag{10}$$

is an orthogonal matrix.

The radar coordinates (range, azimuth, and elevation) of the point P may be obtained immediately in terms of the components of $\mathbf{R}$ in the ENU coordinate system. For if we write

$$\mathbf{R} = x_Q \hat{X}_Q + y_Q \hat{Y}_Q + z_Q \hat{Z}_Q \tag{11}$$

then

$$R = \sqrt{x_Q^2 + y_Q^2 + z_Q^2} \tag{12a}$$

$$A = \arctan \frac{x_Q}{y_Q} \, , \;\; 0 \leq A < 360° \tag{12b}$$

$$E = \arcsin \frac{z_Q}{R} \, , \;\; 0 \leq E \leq 90° \tag{12c}$$

where R, A, and E are respectively range, azimuth, and elevation.

Since

$$\mathbf{r} = \mathbf{r}_Q + \mathbf{R} \tag{13}$$

where $\mathbf{r}_Q$ is the vector from O to Q (see Figs. 2 and 3) eqs. (1) and (5) imply

$$\mathbf{R} = (x - \xi)\hat{X} + (y - \eta)\hat{Y} + (z - \zeta)\hat{Z} \, .$$

Hence from (11) and (9)

$$\begin{bmatrix} x_Q \\ y_Q \\ z_Q \end{bmatrix} = \mathbf{A} \begin{bmatrix} x - \xi \\ y - \eta \\ z - \zeta \end{bmatrix} . \tag{14}$$

Using (4) we may write (14) more explicitly as

$$\begin{bmatrix} x_Q \\ y_Q \\ z_Q \end{bmatrix} = \mathbf{A} \begin{bmatrix} x - (r_F \cos \lambda + h \cos \mu)\cos \ell \\ y - (r_F \cos \lambda + h \cos \mu)\sin \ell \\ z - (r_F \sin \lambda + h \sin \mu) \end{bmatrix}$$

where r_F is given by (2) of Section 1. Thus we may write the vector of range, azimuth, and elevation as a vector function of ECI coordinates and fixed parameters. In the next section the explicit form for the vector radar measurement model shall be presented.

3. THE KALMAN MEASUREMENT MODEL EQUATIONS

In Section 8 of Chapter I we wrote the measurement model for the nonlinear case in the form

$$\mathbf{Z}_k = \mathbf{h}_k(\mathbf{x}(t_k)) + \boldsymbol{v}_k \ , \quad k = 1, 2, \cdots, N \tag{1}$$

where $\mathbf{x}$ was the state vector and $\boldsymbol{v}$ the measurement noise. The coordinate transformations which express the radar observations on range, azimuth and elevation in terms of ECI coordinates and fixed parameters may be cast in this form, namely,

$$\mathbf{h}_k'(\mathbf{x}(t_k)) = \{R(\mathbf{x}(t_k)), A(\mathbf{x}(t_k)), E(\mathbf{x}(t_k))\} \ .$$

We note that $\mathbf{h}_k$ depends also on the parameters ℓ_0, λ, and h, the radar's longitude, geocentric latitude, and height. The components of $\mathbf{h}_k$ are provided by (12) of Section 2.

We usually assume that $\boldsymbol{v}$ belongs to a mean zero vector stochastic process with diagonal covariance matrix. That is, the noise errors on R, A, E are usually assumed to be independent. There are, however, applications where biases and correlated measurement noises are modeled explicitly.

As a nontrivial example of biases and correlated noise, we consider the case of monopulse radar measurements [44]. Suppose that with a radar located at Q we wish to measure the azimuth and elevation of a vehicle in flight. Let the center of mass of the vehicle be at P, while the radar, due to pointing errors, measures values of azimuth and elevation as if the vehicle's center of mass were at P^* (see Fig. 4). Then we may write

$$A = A^* + \delta A \tag{2}$$
$$E = E^* + \delta E$$

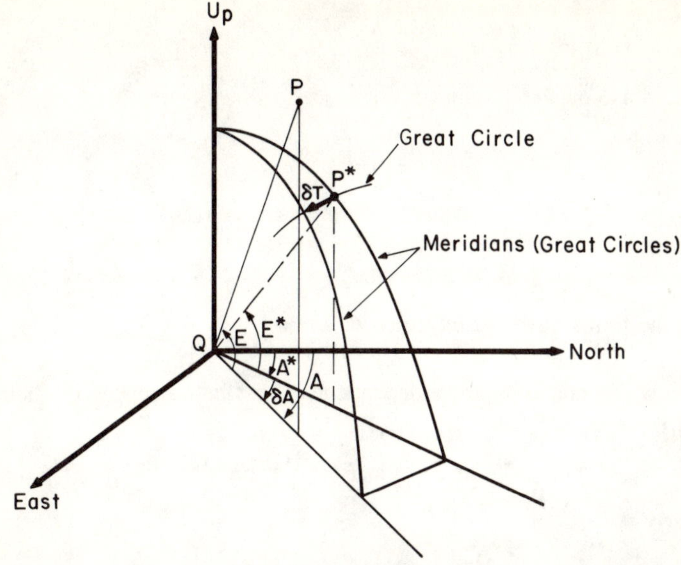

Figure 4
Relation of Azimuth and Traverse

and

$$\delta A = \frac{\delta T}{\cos E^*}$$

where A and E are azimuth and elevation, while T is traverse. The error angles δE and δT also depend on the physical characteristics of the radar.

What appears in the Kalman measurement model is the measured values of A and E. However, what we actually measure with a monopulse radar is A^*, E^* and δA, δE. The first pair corresponds to the *pointing coordinates* of the radar's beam. The latter are small angular *offsets* from the center of the beam to the target. Let us therefore define the measured value of A as

$$A_{\text{meas}} = A_{\text{true}} + b_A + v_A$$

$$= \left[(A^*)_{\text{true}} + \frac{(\delta T)_{\text{true}}}{\cos E^*} \right] + \left(b_{A^*} + \frac{b_{\delta T}}{\cos E^*} \right) + \left(v_{A^*} + \frac{v_{\delta T}}{\cos E^*} \right)$$

where the b's represent measurement biases. Then replacing $\cos E^*$ by $\cos E$, the bias on the azimuth is, approximately,

$$b_A = b_{A*} + \frac{b_{\delta T}}{\cos E} \tag{3}$$

and the noise on the azimuth is

$$v_A = v_{A*} + \frac{v_{\delta T}}{\cos E} \; . \tag{4}$$

Similarly

$$E_{\text{meas}} = E_{\text{true}} + b_E + v_E$$

where the terms have the expected definition.

Now $E*$ and $\delta E*$ are independent. Thus the variances of their respective noises add [see (2)]

$$\text{Var } E = \text{Var } E* + \text{Var } \delta E \; .$$

Similarly

$$\text{Var } A = \text{Var } A* + \text{Var } \delta A$$

$$= \text{Var } A* + \frac{\text{Var } \delta T}{\cos^2 E} \; .$$

It is assumed that in the monopulse processing, only δA and δE are correlated. Thus

$$\text{Cov}(A,E) = \text{Cov}(\delta A, \delta E) = \frac{1}{\cos E} \, \text{Cov}(\delta T, \delta E) \; .$$

In this case the covariance matrix of the radar measurement model is of the form

$$\mathcal{E}vv' = \begin{bmatrix} \text{Var } R & 0 & 0 \\ 0 & \text{Var } A & \text{Cov } (A,E) \\ 0 & \text{Cov } (A,E) & \text{Var } E \end{bmatrix}$$

where Var R is the variance of the noise on range.

We observe in (3) and (4) that both azimuth and traverse contribute to measurement error. For example, if $b_{A*} = 0 = \text{Var } A*$, we may attribute all the error in azimuth to traverse.

If we choose to explicitly include bias in the Kalman measurement model, then (1) must be modified. Let

$$\mathbf{b}' = \{b_R, \, b_A, \, b_E\}$$

be the bias vector for a monopulse radar [where b_A is given by (3) and b_R is the bias on range.] Then we replace (1) by

$$\mathbf{Z}_k = \mathbf{h}_k(\mathbf{x}(t_k)) + \mathbf{b}(t_k) + \mathbf{v}_k \; .$$

If we choose to estimate the bias along with the state vector, then the System model must be augmented to include the bias model,

$$\dot{\mathbf{b}}(t) = \mathbf{0} + \mathbf{noise}$$

and the state vector will be of the form

$$\xi = \begin{bmatrix} \mathbf{x} \\ \mathbf{b} \end{bmatrix} \; .$$

4. ON-BOARD MEASUREMENTS

In Section 2 of Chapter II we derived the equations of motion for the orientation and rotation of a rigid body. The Kalman System model for rigid body dynamics was presented in Section 4 of Chapter II. It may have occurred to the reader that the determination of the six degrees of freedom state vector, from radar data alone, may not be possible. In applications concerning the estimation of a reentry vehicle's orientation and rotation, however, one is sometimes provided with on-board observations of the vehicle's acceleration and rotation. We shall assume that the on-board measurements include direct measurements of the linear accelerations in the $\mathbf{R}_{OL}$, $\mathbf{P}_{IT}$, and $\mathbf{Y}_{AW}$ directions (which are body fixed, see Section 2 of Chapter II) as well as angular velocities about these axes, such as would arise from an idealized on-board measurement system. (For more on on-board instrumentation see [3, 15, 18, 40].)

From (1) of Section 2 of Chapter II

$$\begin{bmatrix} \hat{R}_{OL} \\ \hat{P}_{IT} \\ \hat{Y}_{AW} \end{bmatrix} = \mathbf{E} \begin{bmatrix} \hat{X}_b \\ \hat{Y}_b \\ \hat{Z}_b \end{bmatrix}$$

where $\mathbf{X}_b$, $\mathbf{Y}_b$, $\mathbf{Z}_b$ is the body-centered inertial (BCI) coordinate frame parallel to $\mathbf{X}$, $\mathbf{Y}$, $\mathbf{Z}$, respectively, and $\mathbf{E}$ is the Euler angle transformation. Since $\mathbf{E}$ is an orthogonal matrix, then by the Lemma of Section 1 of Chapter II,

$$\begin{bmatrix} \dot{u}_R \\ \dot{v}_P \\ \dot{w}_Y \end{bmatrix} = \frac{d^2}{dt^2}\, \mathbf{E} \begin{bmatrix} x \\ y \\ z \end{bmatrix}$$

where $\dot{u}_R$, $\dot{v}_P$, $\dot{w}_Y$ are the linear accelerations along the vehicle's principal axes (VPA), $\mathbf{R}_{OL}$, $\mathbf{P}_{IT}$, $\mathbf{Y}_{AW}$, respectively. Thus

$$\begin{bmatrix} \dot{u}_R \\ \dot{v}_P \\ \dot{w}_Y \end{bmatrix} = \ddot{\mathbf{E}} \begin{bmatrix} x \\ y \\ z \end{bmatrix} + 2\dot{\mathbf{E}} \begin{bmatrix} u \\ v \\ w \end{bmatrix} + \mathbf{E} \begin{bmatrix} \dot{u} \\ \dot{v} \\ \dot{w} \end{bmatrix} \tag{1}$$

and from (11) of Section 1 of Chapter II,

$$\begin{bmatrix} \dot{u} \\ \dot{v} \\ \dot{w} \end{bmatrix} = \begin{bmatrix} g_X \\ g_Y \\ g_Z \end{bmatrix} + \frac{Aq}{m}\, \mathbf{D}' \begin{bmatrix} -C_D \\ C_S \\ C_L \end{bmatrix} .$$

Introducing a mean zero measurement noise process $\mathbf{V}$, we may write the measurement model corresponding to (1) succinctly as

$$\mathbf{a}(t) = \mathbf{\Gamma}(\mathbf{x}(t)) + \mathbf{V}(t) \tag{2}$$

where the state vector $\mathbf{x}$ is given by (4) of Section 4 of Chapter II.

If $\mathbf{\Omega}$ is the observation vector on the angular velocities with respect to the VPA coordinate frame, then

$$\mathbf{\Omega}(t) = \mathbf{\omega}(t) + \mathbf{V}^*(t) \tag{3}$$

where $\mathbf{\omega}'(t) = \{\omega_R(t), \omega_P(t), \omega_Y(t)\}$ is the angular velocity vector with respect to the VPA frame, and $\mathbf{V}^*$ is assumed to have mean zero. In a more uniform notation we may write (3) as

$$\mathbf{\Omega}(t) = \mathbf{\Gamma}^*(\mathbf{x}(t)) + \mathbf{V}^*(t) . \tag{4}$$

Biases also may be associated with (2) and (4). In realistic models the bias terms may be needed to account for measurement errors arising from instrumentation misalignment from the VPA frame, or offset from the vehicle's center of mass (see [15, 40]).

Chapter IV

Practical Aspects of Kalman Estimation

0. INTRODUCTION

In this chapter we are concerned with the practical implementation of the Kalman estimation equations. We consider both the filtering and smoothing cases. We shall derive forms for the estimation equations which are particularly suited to numerical computation.

The basic Kalman equations were derived in Chapter I for the linear case, and in Section 8 of Chapter I we presented those for the nonlinear case. In Section 1 of this chapter we summarize these equations and combine them with the results of Chapters II and III. In order to present the main ideas unencumbered by peripheral complications, we use a spherical, rotating earth model, an exponential atmospheric density profile, and drag only for aerodynamic effects.

In Section 2 we discuss the difficulties often encountered with the numerical integration of the covariance matrix differential equation. We derive some approximate approaches to numerical integration and provide some particularly useful equations for propagating covariances.

In Section 3 the smoother is analyzed from the standpoint of numerical accuracy, computational efficiency, and data storage. We conclude this discussion in Section 4 where a practical forward-backward filter and smoother algorithm is given.

Finally, in Section 5 a nontrivial example of smoothing is presented for the problem of estimation of a reentry vehicle's state vector based on radar observations.

1. THE BASIC KALMAN FILTER FOR REENTRY

In Section 8 of Chapter I we discussed the EKF together with the equations for propagating and updating the state estimate vector and covariance matrix. These equations now will be reconsidered in the context of a specific reentry system model and radar measurement model.

We shall take

$$\hat{\mathbf{x}}' = \{\hat{x}, \hat{y}, \hat{z}, \hat{u}, \hat{v}, \hat{w}, \hat{C}_D\}$$

(see Section 4 of Chapter II) as our state estimate vector. The form of the vector equation $\dot{\hat{\mathbf{x}}}(t) = \mathbf{f}(\hat{\mathbf{x}}(t), t)$ for propagating the state estimate (assuming a spherical earth) is

$$
\begin{bmatrix}
\dot{x} \\
\dot{y} \\
\dot{z} \\
\dot{u} \\
\dot{v} \\
\dot{w} \\
\dot{C}_D
\end{bmatrix}
=
\begin{bmatrix}
u \\
v \\
w \\
-\dfrac{GM}{r^3}x - \dfrac{q}{\beta}(u + \omega_e y) \\
-\dfrac{GM}{r^3}y - \dfrac{q}{\beta}(v - \omega_e x) \\
-\dfrac{GM}{r^3}z - \dfrac{q}{\beta}(w) \\
0
\end{bmatrix}
\tag{1}
$$

evaluated at $\mathbf{x} = \hat{\mathbf{x}}$, where $q = \frac{1}{2}\rho|\mathbf{V}_a|^2$ is the magnitude of the dynamic pressure and $\beta = m/AC_D$ is the ballistic coefficient. [See (14) of Section 1 of Chapter II and (2) and (3) of Section 4 of Chapter II.]

The radar will provide observations on the position of the target. The range, azimuth, and elevation are given by (12) of Section 2 of Chapter III. If the radar is located on a spherical earth then $r_F = r_e$, $\mu = \lambda$, $h = 0$, and $e = 0$. Hence if the radar is at geocentric latitude λ and east longitude ℓ_0, then at time t we may write R, A, and E explicitly as

$$R = \sqrt{x^2 + y^2 + z^2 + r_e^2 - 2r_e(x\cos\lambda\cos\ell + y\cos\lambda\sin\ell + z\sin\lambda)} \tag{2a}$$

$$A = \arctan\left(\frac{-x\sin\ell + y\cos\ell}{-x\sin\lambda\cos\ell - y\sin\lambda\sin\ell + z\cos\lambda}\right), \tag{2b}$$

$$0 \leqq A < 2\pi$$

$$E = \arcsin\left(\frac{x \cos \lambda \cos \ell + y \cos \lambda \sin \ell + z \sin \lambda - r_e}{R}\right) , \qquad (2c)$$

$$0 \leqq E \leqq \tfrac{1}{2}\pi$$

where

$$\ell = \ell_0 + \omega_e t .$$

For purposes of numerical computation we must provide an explicit form for the atmospheric mass density ρ. We shall assume that it decreases exponentially with increasing altitude [33]. Thus we may write

$$\rho(r) = \rho_0 e^{-(r-r_e)/h_0}$$

where $r = (x^2 + y^2 + z^2)^{\frac{1}{2}}$. It suffices for our purposes to define ρ_0 and h_0 as

$$\rho_0 = 1.225 \text{ kg}/\text{m}^3$$

$$h_0 = 6,700 \text{ m} .$$

Equations (1) and (2) may be written more compactly as

$$\dot{\hat{\mathbf{x}}} = \mathbf{f}(\hat{\mathbf{x}}(t)) , \quad t_{k-1} \leqq t < t_k$$

and

$$\begin{bmatrix} R(t_k) \\ A(t_k) \\ E(t_k) \end{bmatrix} = \mathbf{h}_k(\mathbf{x}(t_k)) , \quad k = 1, 2, \cdots, N$$

respectively where we have included the explicit dependence on time. Note that for our reentry model $\mathbf{f}$ is a function only of $\hat{\mathbf{x}}(t)$ and not explicitly a function of t.

We now describe the basic estimation procedure. First we must be able to precompute an initial estimate $\hat{\mathbf{x}}_0$ of the vehicle's state vector. We associate with $\hat{\mathbf{x}}_0$ the positive definite matrix P_0. The values of the entries of P_0 are chosen to express the relative weights we choose to assign to the corresponding components of $\hat{\mathbf{x}}_0$. Usually $\hat{\mathbf{x}}_0$ is computed *a priori* based on preprocessing some data; and P_0 is chosen as diagonal, but large, so as not to bias the subsequent processing away from the data.

Next we use $\hat{\mathbf{x}}_0$ and P_0 as initial conditions to the propagation equations. These equations are then integrated to obtain predictions at time t_1:

$$\hat{\mathbf{x}}(t_1^-) = \hat{\mathbf{x}}_0 + \int_{t_0}^{t_1} \dot{\hat{\mathbf{x}}}(t)dt \qquad (3)$$

$$P(t_1^-) = P_0 + \int_{t_0}^{t_1} \dot{P}(t)\,dt \tag{4}$$

where (see Section 8 of Chapter I)

$$\dot{\hat{\mathbf{x}}}(t) = \mathbf{f}(\hat{\mathbf{x}}(t))$$

$$\dot{P}(t) = F(\hat{\mathbf{x}}(t))P(t) + P(t)F'(\hat{\mathbf{x}}(t)) + \Omega(t) \ .$$

We note that we do not have available closed form solutions for (3) and (4). Thus we must perform the integrations numerically and obtain approximations. We shall denote the numerical approximations to $\hat{\mathbf{x}}(t_1^-)$ and $P(t_1^-)$ by $\hat{\mathbf{x}}_1(-)$ and $P_1(-)$ respectively.

Now at time t_1 we have the observation vector $\mathbf{Z}_1$. We therefore use the update equations to operate on $\hat{\mathbf{x}}_1(-)$ and $P_1(-)$ to obtain

$$\hat{\mathbf{x}}_1 = \hat{\mathbf{x}}_1(-) + K_1[\mathbf{Z}_1 - \mathbf{h}_1(\hat{\mathbf{x}}_1(-))]$$

$$P_1 = [I - K_1 H_1(\hat{\mathbf{x}}_1(-))]P_1(-)$$

where

$$K_1 = P_1(-)H_1'(\hat{\mathbf{x}}_1(-))[H_1(\hat{\mathbf{x}}_1(-))P_1(-)H_1'(\hat{\mathbf{x}}_1(-)) + R_1]^{-1} \ ,$$

and the positive definite matrix R_1 represents the assumed covariance matrix of the measurement noise.

We now use $\hat{\mathbf{x}}_1$ and P_1 as new initial conditions to the propagation equations and iterate the process of propagation and updating until the data are exhausted. In general, then, we have for $k = 1, 2, \cdots, N$, the following recursions on $\hat{\mathbf{x}}$ and P:

Numerical propagation

$$\hat{\mathbf{x}}_k(-) = \hat{\mathbf{x}}_{k-1} + \int_{t_{k-1}}^{t_k} \dot{\hat{\mathbf{x}}}(t)\,dt$$

$$P_k(-) = P_{k-1} + \int_{t_{k-1}}^{t_k} \dot{P}(t)\,dt$$

Numerical update

$$\hat{\mathbf{x}}_k = \hat{\mathbf{x}}_k(-) + K_k[\mathbf{Z}_k - \mathbf{h}_k(\hat{\mathbf{x}}_k(-))]$$

$$P_k = [I - K_k H_k(\hat{\mathbf{x}}_k(-))]P_k(-) \ .$$

This algorithm is illustrated in Fig. 1.

We remark that the $\hat{\mathbf{x}}_k(-)$ and $\hat{\mathbf{x}}_k$, $k = 1, 2, \cdots, N$, represent (numerical) approximations to a realization of the optimal Kalman estimates $\hat{\mathbf{x}}(t_k)$. The covariance matrices $P(t_k)$ are being represented by the (numerical) matrices $P_k(-)$ and P_k. These latter "P's" are not true covariances, but rather play the role of weights that express numerically the confidence that should be attributed to their respective numerical "$\hat{\mathbf{x}}$'s". The numerical details of covariance matrix propagation are discussed further in the next section. The numerical significance of the system noise spectral density matrix Ω also is explained there.

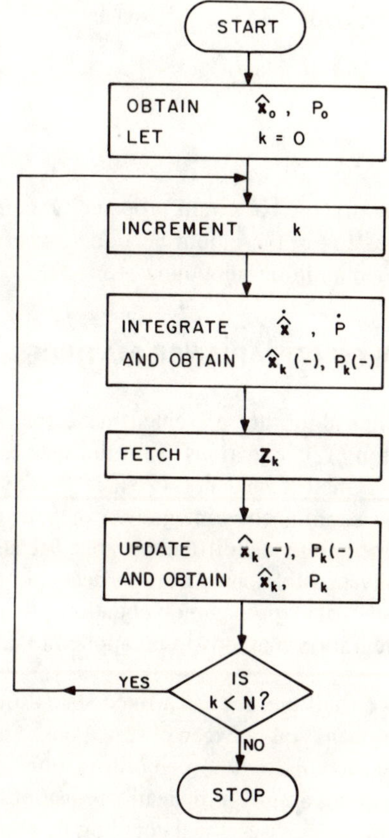

Figure 1
Basic Filter Algorithm

We conclude this section by providing the equations for propagating and updating the corresponding numerical backward filter quantities $\hat{\mathbf{y}}$ and Q, with $k = N-1, N-2, \cdots, 1$ (see Section 4 of Chapter I):

Numerical propagation (backward)

$$\hat{\mathbf{y}}_k(+) = \hat{\mathbf{y}}_{k+1} + \int_{t_{k+1}}^{t_k} \dot{\hat{\mathbf{y}}}(t)\,dt$$

$$Q_k(+) = Q_{k+1} + \int_{t_{k+1}}^{t_k} \dot{Q}(t)\,dt$$

Numerical update (backward)

$$\hat{\mathbf{y}}_k = \hat{\mathbf{y}}(+) + L_k[\mathbf{Z}_k - \mathbf{h}_k(\hat{\mathbf{y}}_k(+))]$$

$$Q_k = [I - L_k H_k(\hat{\mathbf{y}}_k(+))]Q_k(+)$$

where

$$L_k = Q_k(+)H_k'(\hat{\mathbf{y}}_k(+))[H_k(\hat{\mathbf{y}}_k(+))Q_k(+)H_k'(\hat{\mathbf{y}}_k(+)) + R_k]^{-1}.$$

The initial conditions for the backward propagation equations are $\hat{\mathbf{y}}_N$ and Q_N. We defer a discussion of these until Section 4, where we shall discuss the practical implementation of smoothing.

2. PROPAGATION OF COVARIANCE MATRICES

The Kalman filter algorithm presented in Section 1 required the simultaneous integration of 35 equations: 7 for the state estimate vector and 28 for the covariance matrix (and not 49 since P is symmetric). Often, when attempting the simultaneous integration of both the state estimate and covariance matrix, numerical difficulties arise because of system stiffness [9]. It is usually possible, however, to decouple the state estimate vector and covariance matrix propagation equations, thereby allowing different numerical integration methods to be applied to each.

The methods we shall present use a fixed state estimate vector in the covariance matrix propagation between observations. The state estimate is allowed to change at each update between propagations. We thus decouple the state estimate and covariance propagation equations locally, between updates. We maintain piecewise global coupling by allowing the value of the state estimate vector, used to propagate the covariance matrix, to change when we update at each observation.

We fix these ideas. Consider the nonlinear System model

$$\dot{\mathbf{X}}(t) = \mathbf{f}(\mathbf{X}(t)) + \mathbf{V}(t) \tag{1}$$

$$\mathbf{X}(t_k) = \hat{\mathbf{x}}_k$$

in the intervals $[t_k, t_{k+1})$, $k = 0, 1, \cdots, N-1$, where $\mathcal{E}\mathbf{V}(t) = \mathbf{0}$, $\mathcal{E}\mathbf{V}(t)\mathbf{V}'(s) = S(t,s)$, $\mathcal{E}\hat{\mathbf{x}}_k = \mathbf{x}_k$, with the usual assumptions on the independence of $\mathbf{V}$ and $\hat{\mathbf{x}}_k$. We assume we have the state estimates $\hat{\mathbf{x}}(t_k^*)$ where t_k^* is an arbitrary but fixed point in $[t_k, t_{k+1})$, $k = 0, 1, \cdots, N-1$.

Now in Section 8 of Chapter I we defined the matrix F as

$$F(\hat{\mathbf{x}}(t), t) = \left. \frac{\partial \mathbf{f}(\mathbf{x}(t), t)}{\partial \mathbf{x}'(t)} \right|_{\mathbf{x}(t) = \hat{\mathbf{x}}(t)} .$$

If we evaluate $\hat{\mathbf{x}}$ at t_k^* where t_k^* is in the interval $[t_k, t_{k+1})$, then we may define

$$F_k(t) \equiv F(\hat{\mathbf{x}}(t_k^*), t) . \tag{2}$$

Thus, when we locally decouple the covariance matrix and state estimate propagation equations, we may consider the matrix Riccati equation of the form

$$\dot{P}(t) = F_k(t)P(t) + P(t)F_k'(t) + \Omega(t) , \; t_k \leqq t < t_{k+1}$$
$$P(t_k) = P_k , \; k = 0, 1, \cdots, N-1 \tag{3}$$

where Ω is a spectral density matrix,

$$S(t,s) = \Omega(t)\delta(t-s) .$$

The matrix P, in this context, is an approximation to the covariance matrix of $\hat{\mathbf{x}}$ (see, again, Section 8 of Chapter I).

The rationale for using this approximation for the covariance matrix is that the system noise spectral density matrix is usually, in practice, numerically adjusted to compensate for unmodeled errors (both known and unknown) in the System model, and also to compensate for effects from limited precision in computer arithmetic. The process of empirically adjusting the spectral density matrix is sometimes described as "tuning" the filter.

Let $W(t,s)$ be the one-sided Green's function matrix associated with $F(t)$. (We shall drop the subscript k on F for the rest of this section to simplify the notation.) From Section 2 of Chapter I we have

$$P(t) = W(t, t_k)P_k W'(t, t_k) + \int_{t_k}^{t} W(t,s)\Omega(s)W'(t,s)ds . \tag{4}$$

Now (4) represents an *exact* equation for the computation of $P(t)$ from (3). If we wish to calculate $P(t_{k+1}^-)$ where $\tau = t_{k+1} - t_k > 0$ is small, we may consider approximations. Let $F(t) = F$ and $\Omega(t) = \Omega$ be independent of t in this interval. This assumption simplifies the arithmetic, although the methods we shall describe may be carried through even if F and/or Ω is nonconstant. For a constant F,

$$W(t,s) = e^{F(t-s)} . \tag{5}$$

a. Linear approximation (Method 1)

Let $t = t_{k+1}^-$ and approximate the integral in (4) by the integrand evaluated at the lower limit times the length of the interval of integration. Then we obtain

$$P_k(-) = e^{\tau F} P_k e^{\tau F'} + \tau e^{\tau F} \Omega e^{\tau F'} \tag{6}$$
$$= e^{\tau F}(P_k + \tau \Omega)e^{\tau F'} .$$

Now we may write

$$e^{\tau F} = I + \tau F + \tfrac{1}{2}\tau^2 F^2 + \cdots . \tag{7}$$

Thus if we replace $e^{\tau F}$ in (6) by the linear approximation

$$\Phi(\tau) = I + \tau F ,$$

then (6) becomes

$$[P_k(-)]_{\text{LIN}-1} = \Phi(\tau)(P_k + \tau\Omega)\Phi'(\tau)$$
$$= P_k + \tau(FP_k + P_k F' + \Omega) + \tau^2(FP_k F' + F\Omega + \Omega F') \tag{8}$$
$$+ \tau^3 F\Omega F' .$$

b. Linear approximation (Method 2)

We may write

$$P(t_{k+1}^-) = P(t_k) + \tau\dot{P}(t_k) + \tfrac{1}{2}\tau^2 \ddot{P}(t_k) + \cdots . \tag{9}$$

Hence

$$P_k(-) = P(t_k) + \tau\dot{P}(t_k) \tag{10}$$

also may be considered as a first order approximation. But from (4)

$$\dot{P}(t) = FP(t) + P(t)F' + \Omega . \tag{11}$$

Therefore we may write (10) as

$$[P_k(-)]_{\text{LIN}-2} = P_k + \tau(FP_k + P_k F' + \Omega) \tag{12}$$

—which agrees, to order τ, with (8).

c. Higher order approximations

Method 2 generalizes to higher orders of approximation. Method 1 does not. From (11)

$$\ddot{P}(t) = [F^2 P(t) + 2FP(t)F' + P(t)F'^2] + [F\Omega + \Omega F']$$

$$\dddot{P}(t) = [F^3 P(t) + 3F^2 P(t)F' + 3FP(t)F'^2 + P(t)F'^3]$$

$$+ [F^2\Omega + 2F\Omega F' + \Omega F'^2] \ ,$$

and if desired, higher order derivatives of $P(t)$ may be calculated. In particular we write out the quadratic and tertiary approximations to $P_k(-)$.

d. Quadratic approximation
From (9)

$$[P_k(-)]_{\text{QUAD}} = P_k + \tau \dot{P}_k + \tfrac{1}{2}\tau^2 \ddot{P}_k$$

$$= P_k + \tau(FP_k + P_k F' + \Omega)$$

$$+ \tfrac{1}{2}\tau^2(F^2 P_k + 2FP_k F' + P_k F'^2 + F\Omega + \Omega F')$$

is a second order approximation. Compare the coefficient of τ^2 in the above equation with that of (8). In particular the coefficients of τ^2 in (a) and (d) both contain the symmetric term $FP_k F'$.

e. Tertiary approximation
As above

$$[P_k(-)]_{\text{TERT}} = [P_k(-)]_{\text{QUAD}} + \tfrac{1}{6}\tau^3(F^3 P_k + 3F^2 P_k F'$$

$$+ 3FP_k F'^2 + P_k F'^3 + F^2\Omega + 2F\Omega F' + \Omega F'^2) \ .$$

In particular (a) contains $F\Omega F'$ as the coefficient of τ^3, while (e) contains $\tfrac{1}{3}F\Omega F'$ as part of the coefficient of τ^3.

It seems appropriate at this point to make a slight digression on matric Riccati equations. Consider then, the matric Riccati equation

$$\dot{X} = A(t)X + XB(t) + XC(t)X + F(t) \tag{13a}$$

where $A(t)$, $B(t)$, $C(t)$, $F(t)$ are $n \times n$ matrices analytic on some t interval T. Together with (13a) we associate the initial condition

$$X(t_0) = X_0 \ , \ t_0 \in T \ . \tag{13b}$$

Now we know that (13) has a unique solution for t in a neighborhood $\mathfrak{N} \subset T$ of t_0. Thus for $t \in \mathfrak{N}$

$$X(t) = X(t_0) + \dot{X}(t_0)(t - t_0) + \tfrac{1}{2}\ddot{X}(t_0)(t - t_0)^2 + \cdots \ .$$

In particular, if $t_1 \in \mathfrak{N}$ and $t_1 - t_0 = \tau > 0$ is small compared with one, a small finite number of the terms in the infinite series

$$X(t_1) = X(t_0) + \tau \dot{X}(t_0) + \tfrac{1}{2}\tau^2 \ddot{X}(t_0) + \cdots$$

should be a good approximation to $X(t_1)$.

Since, from (13a)

$$\ddot{X} = A\dot{X} + \dot{A}X + \dot{X}B + X\dot{B} + \dot{X}CX + XC\dot{X} + XCX + \dot{F} \tag{14}$$

we may substitute $\dot{X}$ as given by (13a) into (14) to obtain $\ddot{X}(t_1)$ as a function only of X (and A, B, C, F) evaluated at $t = t_0$ (similarly for higher derivatives). Thus any matrix satisfying a Riccati equation has the above type of representation. (The method extends, of course, to differential equations other than the Riccati.)

But if P is the covariance matrix of the forward filter, Q the covariance matrix of the backward filter, and M the covariance matrix of the smoother, then P, P^{-1}, Q, Q^{-1}, and M all satisfy Riccati equations, viz.:

$$\dot{P} = FP + PF' + \Omega$$

$$\dot{P}^{-1} = -P^{-1}F - F'P^{-1} - P^{-1}\Omega P^{-1}$$

$$\dot{Q} = FQ + QF' - \Omega$$

$$\dot{Q}^{-1} = -Q^{-1}F - F'Q^{-1} + Q^{-1}\Omega Q^{-1}$$

$$\dot{M} = (F - \Omega Q^{-1})M + M(F - \Omega Q^{-1})' + \Omega \ .$$

(Recall that $\dot{R}^{-1}$ means $\dfrac{d}{dt}R^{-1}(t)$ and not $[\dot{R}(t)]^{-1}$. Also $\dot{M}^{-1}$ satisfies a Riccati equation, although we do not need this fact.) Hence the method we outlined in detail for P is applicable to P^{-1}, Q, Q^{-1}, and M as well.

We further may simplify the equations for propagating the covariance matrix if we change our use of system noise. Define S_k as

$$S_k = \int_{t_k}^{t_{k+1}} W(t_{k+1},s)\Omega(s)W'(t_{k+1},s)ds \ . \tag{15}$$

We note that the arguments used at the beginning of this section on the empirical derivation of the numerical values of Ω also may be applied to S_k. That is, we may use S_k, $k = 0,1,\cdots,N-1$ directly to tune the filter and thus obviate the necessity for the numerical consideration of Ω.

If we assume that S_k, $k = 0,1,\cdots,N-1$ is given, then the differences in approximation, for the methods outlined in this section, arise solely from the truncation of the series expansion of W, see (5) and (7).

From (4) and (15) we have, letting $P_k \equiv P(t_k)$,

$$P(t_{k+1}^-) = W_k P_k W_k' + S_k \tag{16}$$

where

$$W_k = W(t_{k+1}, t_k) .$$

Thus, for the linear approximation case, we have

$$[P_k(-)]_{\text{LIN}} = \Phi(\tau) P_k \Phi'(\tau) + S_k$$

$$= P_k + \tau(FP_k + P_k F') + \tau^2 FP_k F' + S_k .$$

For the general case, using a J-th order expansion for W,

$$[P_k(-)]_J = \left[\sum_{j=0}^{J} \frac{(\tau F)^j}{j!} \right] P_k \left[\sum_{j=0}^{J} \frac{(\tau F)^j}{j!} \right]' + S_k .$$

[We remark that we have tacitly assumed that the t_k are uniformly spaced, that is,

$$\tau = t_{k+1} - t_k , \quad k = 0, 1, \cdots, N-1$$

independent of k. We may accept arbitrarily spaced data (provided $t_k < t_{k+1}$), but then we must use τ_k in the above formulas.]

We also shall need the equation for propagating the backward covariance matrix in a form similar to (16). We recall that the intervals for backward propagation are $t_k < t \leq t_{k+1}$, $k = 0, 1, \cdots, N-1$. From Section 4 of Chapter I (with $B = I$) we have, letting $Q_{k+1} \equiv Q(t_{k+1})$,

$$Q(t_k^+) = W(t_k, t_{k+1}) Q_{k+1} W'(t_k, t_{k+1}) - \int_{t_{k+1}}^{t_k} W(t_k, s) \Omega(s) W'(t_k, s) ds . \tag{17}$$

To relate the integral in (17) to S_k we recall [see (2b) of Section 2 of Chapter I] that

$$W(t_k, s) = W(t_k, t_{k+1}) W(t_{k+1}, s) .$$

Using this result together with (15) yields

$$\int_{t_{k+1}}^{t_k} W(t_k, s) \Omega(s) W'(t_k, s) ds$$

$$= W(t_k, t_{k+1}) \left[\int_{t_{k+1}}^{t_k} W(t_{k+1}, s) \Omega(s) W'(t_{k+1}, s) ds \right] W'(t_k, t_{k+1})$$

$$= - W(t_k, t_{k+1}) S_k W'(t_k, t_{k+1}) .$$

The backward covariance matrix propagation equation is thus

$$Q(t_k^+) = W_k^{-1} [Q_{k+1} + S_k] W_k'^{-1}$$

where we have used the identity

$$W_k^{-1} \equiv W^{-1}(t_{k+1},t_k) = W(t_k,t_{k+1}) \ .$$

Sometimes we need to reverse the direction of propagation. To avoid confusion, we list these propagation equations, together with their corresponding forward and backward forms.

Forward propagation

$$P(t_{k+1}^-) = W_k P_k W_k' + S_k$$

Reverse-forward propagation

$$P_k = W_k^{-1}[P(t_{k+1}^-) - S_k]W_k'^{-1}$$

Backward propagation

$$Q(t_k^+) = W_k^{-1}[Q_{k+1} + S_k]W_k'^{-1}$$

Reverse-backward propagation

$$Q_{k+1} = W_k Q(t_k^+)W_k' - S_k$$

3. ALTERNATIVE FORMS FOR THE SMOOTHER

In Section 5 of Chapter I we derived the equations for the smoother based on separate forward and backward filters. We now would like to develop some equivalent forms commonly used in the literature. Those lacking practical value are nevertheless useful in attacking theoretical problems associated with the Kalman filter.

We recall from Chapter I that the smoothed state estimate $\hat{\mathbf{s}}(t_k) \equiv \hat{\mathbf{s}}_k$ and covariance matrix $M(t_k) \equiv M_k$ are given by

$$\hat{\mathbf{s}}_k = M_k[P^{-1}(t_k^-)\hat{\mathbf{x}}(t_k^-) + Q_k^{-1}\hat{\mathbf{y}}_k]$$

$$= M_k[P_k^{-1}\hat{\mathbf{x}}_k + Q^{-1}(t_k^+)\hat{\mathbf{y}}(t_k^+)] \tag{1}$$

and

$$M_k^{-1} = P^{-1}(t_k^-) + Q_k^{-1}$$

$$= P_k^{-1} + Q^{-1}(t_k^+) \ . \tag{2}$$

The $\hat{\mathbf{x}}(t_k^-)$, $P(t_k^-)$ are predictions by a forward filter, while $\hat{\mathbf{y}}(t_k^+)$, $Q(t_k^+)$ are predictions by a backward filter, with $\hat{\mathbf{x}}_k$, P_k, and $\hat{\mathbf{y}}_k$, Q_k the corresponding updated values respectively.

The form for the smoother equations given by (1) and (2) is known as the Fraser-Potter (F-P) smoother equations [17]. They require three matrix inversions. The summing of terms having the form $P^{-1}\hat{x}$ and $Q^{-1}\hat{y}$, however, can give rise to errors due to the finite precision of the computer. We can avoid these difficulties by reformulating the smoother equations via some useful matrix identities. In general, if M, P, Q are symmetric and nonsingular, then

$$M^{-1} = P^{-1} + Q^{-1} \tag{3}$$

implies

$$M = (P^{-1} + Q^{-1})^{-1}$$

$$= P\,(P+Q)^{-1}Q \tag{4a}$$

$$= Q\,(P+Q)^{-1}P \ . \tag{4b}$$

Also, if

$$\zeta = M(P^{-1}\xi + Q^{-1}\eta)$$

then using (3),

$$\zeta = M[(M^{-1} - Q^{-1})\xi + Q^{-1}\eta]$$

$$= M[P^{-1}\xi + (M^{-1} - P^{-1})\eta] \ ,$$

or, after multiplying through by M and simplifying,

$$\zeta = \xi + MQ^{-1}(\eta - \xi) \tag{5}$$

$$= \eta + MP^{-1}(\xi - \eta) \ .$$

If we further substitute (4) into (5) we obtain

$$\zeta = \xi + P(P+Q)^{-1}(\eta - \xi) \tag{6}$$

$$= \eta + Q(P+Q)^{-1}(\xi - \eta) \ .$$

Thus, if we use updated forward filter and predicted backward filter values, the smoother equations may be written as

$$\hat{s}_k = \hat{y}(t_k^+) + M_k P_k^{-1}[\hat{x}_k - \hat{y}(t_k^+)] \tag{7a}$$

$$= \hat{y}(t_k^+) + Q(t_k^+)[P_k + Q(t_k^+)]^{-1}[\hat{x}_k - \hat{y}(t_k^+)] \ . \tag{7b}$$

Alternate forms of (7) are obtained if we interchange the roles of the forward and backward filters. These shall be presented later in this section. The forms for the smoother equations given in (7) are better conditioned numerically than those of the F-P form (1) since they involve only the residual between two state estimates. The form provided by (7b) has the additional advantage that only one matrix inversion is required (provided knowledge of the smoother covariance matrix is not needed).

We wish to derive another form of the smoother equations. Towards this end consider

$$M_{k+1}^{-1} = P^{-1}(t_{k+1}^-) + Q_{k+1}^{-1} \tag{8}$$

together with

$$M_k^{-1} = P_k^{-1} + Q^{-1}(t_k^+) \ . \tag{9}$$

Then from the propagation equations listed at the end of the previous section,

$$P(t_{k+1}^-) + Q_{k+1} = W_k[P_k + Q(t_k^+)]W_k'$$

or

$$[P_k + Q(t_k^+)]^{-1} = W_k'[P(t_{k+1}^-) + Q_{k+1}]^{-1}W_k \ . \tag{10}$$

From the identity

$$M = P - P(P+Q)^{-1}P \tag{11}$$

evaluated at $t = t_k^+$ and $t = t_{k+1}^-$ we have

$$[P_k + Q(t_k^+)]^{-1} = P_k^{-1} - P_k^{-1}M_kP_k^{-1}$$

and

$$[P(t_{k+1}^-) + Q_{k+1}]^{-1} = P^{-1}(t_{k+1}^-) - P^{-1}(t_{k+1}^-)M_kP^{-1}(t_{k+1}^-) \ .$$

Substitution of these formulas in (10) yields

$$P_k^{-1} - P_k^{-1}M_kP_k^{-1} = W_k'[P^{-1}(t_{k+1}^-) - P^{-1}(t_{k+1}^-)M_kP^{-1}(t_{k+1}^-)]W_k \ , \tag{12}$$

which after simplification, gives the desired propagation equation for M:

$$M_k = P_k + C_k[M_{k+1} - P(t_{k+1}^-)]C_k' \tag{13}$$

where

$$C_k = P_kW_k'P^{-1}(t_{k+1}^-) \ . \tag{14}$$

This form of the smoother propagation equation is sometimes called the Rauch-Tung-Striebel (RTS) propagation equation [41].

An alternate form of the RTS propagation equation which involves Q and Q^{-1} also may be derived. If we use the identity

$$M = Q - Q(P+Q)^{-1}Q \tag{15}$$

[which is (11) with P and Q interchanged] then similar arguments show that

$$M_k = Q(t_k^+) + D_k(M_{k+1} - Q_{k+1})D_k' \tag{16}$$

where

$$D_k = Q(t_k^+)W_k'Q_{k+1}^{-1} \; . \tag{17}$$

We need to specify the initial conditions for the backward filter and smoother propagation equations. We smooth in the interval $[t_1,t_N]$. Although there may be independent information available in (t_N,∞) we may ignore this. Thus

$$Q^{-1}(t_N^+) = 0 \; . \tag{18}$$

This implies [see (2)] that

$$M_N = P_N \; . \tag{19}$$

If we have a linear system model, we also may propagate directly the smoothed state vector. For the linear case we have, from Sections 2, 3, and 4 of Chapter I, but in the notation of this chapter, the following propagation equations.

Forward

$$\hat{\mathbf{x}}(t_{k+1}^-) = W_k \hat{\mathbf{x}}_k \tag{20}$$

where $\hat{\mathbf{x}}_k \equiv \hat{\mathbf{x}}(t_k)$

Backward

$$\hat{\mathbf{y}}(t_k^+) = W_k^{-1}\hat{\mathbf{y}}_{k+1} \tag{21}$$

where $\hat{\mathbf{y}}_{k+1} \equiv \hat{\mathbf{y}}(t_{k+1})$.

If we properly identify the symbols in (5), then (1) yields

$$\hat{\mathbf{s}}_k = \hat{\mathbf{x}}_k + M_k Q^{-1}(t_k^+)[\hat{\mathbf{y}}(t_k^+) - \hat{\mathbf{x}}_k] \tag{22}$$

and

$$\hat{\mathbf{s}}_{k+1} = \hat{\mathbf{x}}(t_{k+1}^-) + M_{k+1}Q_{k+1}^{-1}[\hat{\mathbf{y}}_{k+1} - \hat{\mathbf{x}}(t_{k+1}^-)] \; . \tag{23}$$

But from (20) and (21)

$$\hat{\mathbf{x}}(t_{k+1}^-) - \hat{\mathbf{y}}_{k+1} = W_k[\hat{\mathbf{x}}_k - \hat{\mathbf{y}}(t_k^+)]$$

or

$$Q_{k+1}M_{k+1}^{-1}[\hat{\mathbf{s}}_{k+1} - \hat{\mathbf{x}}(t_{k+1}^-)] = W_k Q(t_k^+)M_k^{-1}(\hat{\mathbf{s}}_k - \hat{\mathbf{x}}_k) \; . \tag{24}$$

Thus

$$\hat{\mathbf{s}}_k = \hat{\mathbf{x}}_k + M_k Q^{-1}(t_k^+)W_k^{-1}Q_{k+1}M_{k+1}^{-1}[\hat{\mathbf{s}}_{k+1} - \hat{\mathbf{x}}(t_{k+1}^-)] \; . \tag{25}$$

This result may be simplified further. From the forward and reverse-backward propagation equations listed at the end of Section 2

$$W_k[P_k + Q(t_k^+)]W_k' = P(t_{k+1}^-) + Q_{k+1} \ . \tag{26}$$

Now multiply by $P^{-1}(t_{k+1}^-)$ on the right and $Q^{-1}(t_k^+)W_k^{-1}$ on the left to obtain

$$Q^{-1}(t_k^+)P_kW_k'P^{-1}(t_{k+1}^-) + W_k'P^{-1}(t_{k+1}^-)$$

$$= Q^{-1}(t_k^+)W_k^{-1} + Q^{-1}(t_k^+)W_k^{-1}Q_{k+1}P^{-1}(t_{k+1}^-) \ .$$

Rearranging terms leads to

$$[Q^{-1}(t_k^+) + P_k^{-1}]P_kW_k'P^{-1}(t_{k+1}^-) = Q^{-1}(t_k^+)W_k^{-1}Q_{k+1}[Q_{k+1}^{-1} + P^{-1}(t_{k+1}^-)]$$

and from (2),

$$M_k^{-1}P_kW_k'P^{-1}(t_{k+1}^-) = Q^{-1}(t_k^+)W_k^{-1}Q_{k+1}M_{k+1}^{-1} \ .$$

Thus from (14)

$$C_k = M_kQ^{-1}(t_k^+)W_k^{-1}Q_{k+1}M_{k+1}^{-1} \tag{27}$$

and (25) reduces to

$$\hat{\mathbf{s}}_k = \hat{\mathbf{x}}_k + C_k[\hat{\mathbf{s}}_{k+1} - \hat{\mathbf{x}}(t_{k+1}^-)] \tag{28}$$

—which is our desired simplification.

Similarly, if we use (7a) and

$$\hat{\mathbf{s}}_{k+1} = \hat{\mathbf{y}}_{k+1} + M_{k+1}P^{-1}(t_{k+1}^-)[\hat{\mathbf{x}}(t_{k+1}^-) - \hat{\mathbf{y}}_{k+1}]$$

then identical arguments yield

$$\hat{\mathbf{s}}_k = \hat{\mathbf{y}}(t_k^+) + D_k(\hat{\mathbf{s}}_{k+1} - \hat{\mathbf{y}}_{k+1}) \tag{29}$$

where [see (17)]

$$D_k = M_kP_k^{-1}W_k^{-1}P(t_{k+1}^-)M_{k+1}^{-1} \ . \tag{30}$$

Equations (28) and (29) are applicable only to linear system models. We also could have derived similar equations for the EKF. However their utility is limited because the expressions are far more cumbersome and only approximate [20]. Additionally, the absence of any explicit updating makes these equations prone to the accumulation of numerical errors. (This latter point affects even the linear case.) Thus, except for some trivial cases, it is usually better to maintain separate and complete forward and backward filter solutions, and to use equations such as (7) for computing the smoothed state estimate.

The RTS equations for propagating the smoother covariance also have the disadvantage that the associated filter covariance matrix and its inverse must be propagated in parallel with M and the backward filter. The P-form, (13) and (14), requires P_k and $P^{-1}(t_{k+1}^-)$. The Q-form, (16) and (17), requires $Q(t_k^+)$ and Q_{k+1}^{-1}. The Q-form suffers from the further disadvantage that consistency with the smoother's covariance matrix requires Q_N^{-1} to be zero; but $Q(t_N^+)$ cannot be singular in the backward filter. (In the linear case this latter objection may be removed by filtering $Q^{-1}\hat{\mathbf{y}}$ directly, instead of Q and $\hat{\mathbf{y}}$ separately [20].)

We present another form for the smoother which does not suffer from these difficulties. First we consider the P-form. Substitution of the reverse-forward propagation equation in (13) and (14) leads to

$$M_k = W_k^{-1}[P(t_{k+1}^-) - S_k]W_k'^{-1} + C_k[M_{k+1} - P(t_{k+1}^-)]C_k'$$

where

$$C_k = W_k^{-1}[I - S_k P^{-1}(t_{k+1}^-)] \ .$$

Combining these expressions we have

$$M_k = W_k^{-1}\{[I - S_k P^{-1}(t_{k+1}^-)]M_{k+1}[I - S_k P^{-1}(t_{k+1}^-)]'$$

$$+ S_k - S_k P^{-1}(t_{k+1}^-)S_k\}W_k'^{-1} \ .$$

Similarly, if we substitute the backward propagation equation in (16) and (17), we obtain

$$M_k = W_k^{-1}[(I + S_k Q_{k+1}^{-1})M_{k+1}(I + S_k Q_{k+1}^{-1})'$$

$$- S_k - S_k Q_{k+1}^{-1}S_k]W_k'^{-1} \ .$$

We now shall make a slight digression and discuss the similarity between the forms of the smoother equations and filter equations, as well as their relation to the BLUE of $\mathbf{x}$ described in Section 1 of Chapter I. The measurement model we shall need is

$$\mathbf{Z} = H\mathbf{x} + \boldsymbol{v} \tag{31}$$

where H has maximal rank, $\mathcal{E}\boldsymbol{v} = \mathbf{0}$, and $\mathcal{E}\boldsymbol{v}\boldsymbol{v}' = R$ (positive definite.)

Let $\hat{\mathbf{x}}(-)$ and $\hat{\mathbf{y}}(+)$ be unbiased estimators of $\mathbf{x}$ with positive definite covariance matrices $P(-)$ and $Q(+)$, respectively. We may introduce the trivial measurement models

$$\hat{\mathbf{x}}(-) = I\mathbf{x} + \tilde{\mathbf{x}}(-) \tag{32}$$

$$\hat{\mathbf{y}}(+)=I\mathbf{x}+\widetilde{\mathbf{y}}(+) \ . \tag{33}$$

From the definitions of $\hat{\mathbf{x}}(-)$ and $\hat{\mathbf{y}}(+)$ it follows that $\widetilde{\mathbf{x}}(-)$ and $\widetilde{\mathbf{y}}(+)$ have mean zero and respective covariance matrices $P(-)$ and $Q(+)$. We assume that $\boldsymbol{\nu}$, $\widetilde{\mathbf{x}}(-)$, and $\widetilde{\mathbf{y}}(+)$ are all mutually independent.

In Chapter I we saw that the BLUE of $\mathbf{x}$ based on $\hat{\mathbf{x}}(-)$, $\hat{\mathbf{y}}(+)$, and $\mathbf{Z}$ was given by

$$\hat{\mathbf{s}}=M[P^{-1}(-)\hat{\mathbf{x}}(-)+Q^{-1}(+)\hat{\mathbf{y}}(+)+H'R^{-1}\mathbf{Z}] \tag{34}$$

where

$$M^{-1}=P^{-1}(-)+Q^{-1}(+)+H'R^{-1}H \ . \tag{35}$$

The linear property of the BL operator allows us to group the terms in (34) and (35) and form intermediate estimates of $\mathbf{x}$. For example, if we associate $\hat{\mathbf{x}}(-)$ with $\mathbf{Z}$, then we may form the estimate

$$\hat{\mathbf{x}}=P[P^{-1}(-)\hat{\mathbf{x}}(-)+H'R^{-1}\mathbf{Z}] \tag{36}$$

where

$$P^{-1}=P^{-1}(-)+H'R^{-1}H \ . \tag{37}$$

Then (34) and (35) may be written as

$$\hat{\mathbf{s}}=M[P^{-1}\hat{\mathbf{x}}+Q^{-1}(+)\hat{\mathbf{y}}(+)] \tag{38}$$

where

$$M^{-1}=P^{-1}+Q^{-1}(+) \ . \tag{39}$$

Similarly, if we associate $\hat{\mathbf{y}}(+)$ with $\mathbf{Z}$, then

$$\hat{\mathbf{y}}=Q[Q^{-1}(+)\hat{\mathbf{y}}(+)+H'R^{-1}\mathbf{Z}] \tag{40}$$

where

$$Q^{-1}=Q^{-1}(+)+H'R^{-1}H \tag{41}$$

and

$$\hat{\mathbf{s}}=M[P^{-1}(-)\hat{\mathbf{x}}(-)+Q^{-1}\hat{\mathbf{y}}] \tag{42}$$

where

$$M^{-1}=P^{-1}(-)+Q^{-1} \ . \tag{43}$$

Observe that $\hat{\mathbf{x}}$ and $\hat{\mathbf{y}}$ are not independent since they both make use of $\mathbf{Z}$.

We may cast (36) and (40) in the Kalman update form as follows. Substitution of (37) into (36) yields, after simplifying,

$$\hat{\mathbf{x}} = \hat{\mathbf{x}}(-) + PH'R^{-1}[\mathbf{Z} - H\hat{\mathbf{x}}(-)]$$

$$\equiv \hat{\mathbf{x}}(-) + K[\mathbf{Z} - H\hat{\mathbf{x}}(-)] \ , \tag{44}$$

while substitution of (41) into (40) yields

$$\hat{\mathbf{y}} = \hat{\mathbf{y}}(+) + QH'R^{-1}[\mathbf{Z} - H\hat{\mathbf{y}}(+)]$$

$$\equiv \hat{\mathbf{y}}(+) + L[\mathbf{Z} - H\hat{\mathbf{y}}(+)] \ , \tag{45}$$

where K and L are the respective gain matrices. [See (6) of Section 3 and (6) of Section 4 in Chapter I.]

The smoother also may be cast into the form of the update equation. Define

$$\hat{\mathbf{s}}(*) = M(*)[P^{-1}(-)\hat{\mathbf{x}}(-) + Q^{-1}(+)\hat{\mathbf{y}}(+)] \tag{46}$$

where

$$M^{-1}(*) = P^{-1}(-) + Q^{-1}(+) \ . \tag{47}$$

We then may write (34) and (35) as

$$\hat{\mathbf{s}} = M[M^{-1}(*)\hat{\mathbf{s}}(*) + H'R^{-1}\mathbf{Z}] \tag{48}$$

where

$$M^{-1} = M^{-1}(*) + H'R^{-1}H \ . \tag{49}$$

Substitution of (49) into (48) yields, as before,

$$\hat{\mathbf{s}} = \hat{\mathbf{s}}(*) + MH'R^{-1}[\mathbf{Z} - H\hat{\mathbf{s}}(*)]$$

$$= \hat{\mathbf{s}}(*) + G[\mathbf{Z} - H\hat{\mathbf{s}}(*)] \tag{50}$$

where $G = MH'R^{-1}$ is a gain matrix.

The F-P smoother equations also may be put into an update form, for, from (38) and (39)

$$\hat{\mathbf{s}} = \hat{\mathbf{x}} + MIQ^{-1}(+)[\hat{\mathbf{y}}(+) - I\hat{\mathbf{x}}] \tag{51a}$$

or

$$\hat{\mathbf{s}} = \hat{\mathbf{y}}(+) + MIP^{-1}[\hat{\mathbf{x}} - I\hat{\mathbf{y}}(+)] \ . \tag{52a}$$

Similarly, from (42) and (43),

$$\hat{\mathbf{s}} = \hat{\mathbf{x}}(-) + MIQ^{-1}[\hat{\mathbf{y}} - I\hat{\mathbf{x}}(-)] \tag{51b}$$

or

$$\hat{\mathbf{s}} = \hat{\mathbf{y}} + MIP^{-1}(-)[\hat{\mathbf{x}}(-) - I\hat{\mathbf{y}}] \ . \tag{52b}$$

We have used the identity matrix I in the above formulas to imply the underlying role of the measurement model relations (32) and (33), and hence to emphasize the interpretation of (51) and (52) as updating. In (51) and (52) the matrices $MIQ^{-1}(+)$, MIP^{-1}, MIQ^{-1}, $MIP^{-1}(-)$ play the role of gain matrices.

4. PRACTICAL SMOOTHER ALGORITHMS

In this section we summarize the various smoother equations and evaluate their use from a computer processing and storage requirements standpoint. In any of the smoother algorithms considered here, we must first run the data through a forward filter and store intermediate estimates and, possibly, covariance matrices. Thus we use the algorithm presented in Section 1 as our starting point.

Let our initial state estimate, $\hat{\mathbf{x}}_0$, together with its (positive definite) covariance matrix P_0 be given. The data shall consist of N observation vectors, $\mathbf{Z}_k$, $k=1,2,\cdots,N$. For convenience let P_0 be diagonal. Thus P_0^{-1} is obtained trivially. For the discussion which follows we shall use a second order approximation to propagate the state vector. (For more precise methods consult [9].)

We propagate $\hat{\mathbf{x}}$ in two stages. (The reason for a two step procedure will be made clear shortly.) Let

$$t_{k+\frac{1}{2}}=\tfrac{1}{2}(t_{k+1}+t_k) \tag{1}$$

for each k, $k=0,1,\cdots,N-1$. That is, $t_{k+\frac{1}{2}}$ is the midpoint of the interval $[t_k,t_{k+1})$. In Section 2 we presented the formula for a second order approximation to the covariance matrix propagation. The corresponding equation for the state estimate is obtained similarly.

Consider the vector equation

$$\dot{\mathbf{x}}(t)=\mathbf{f}(\mathbf{x}(t)) \ , \ t \in T \tag{2a}$$

where $\mathbf{f}$ is a smooth function. Together with (2a) we associate the initial condition

$$\mathbf{x}(t_0)=\mathbf{x}_0 \ , \ t_0 \in T \ . \tag{2b}$$

Now we know that (2) has a unique solution for τ sufficiently small where $t=t_0+\tau$. Thus if we write

$$\mathbf{x}(t)=\mathbf{x}(t_0)+\dot{\mathbf{x}}(t_0)\tau+\tfrac{1}{2}\ddot{\mathbf{x}}(t_0)\tau^2+\cdots,$$

the derivatives are given by

$$\dot{\mathbf{x}}(t_0) = \mathbf{f}(\mathbf{x}(t_0))$$

and

$$\ddot{\mathbf{x}}(t_0) = \frac{d\mathbf{f}(\mathbf{x}(t))}{dt}\bigg|_{t=t_0}$$

$$= \frac{d\mathbf{f}(\mathbf{x}(t))}{d\mathbf{x}'(t)}\frac{d\mathbf{x}(t)}{dt}\bigg|_{\mathbf{x}(t)=\mathbf{x}(t_0),\ t=t_0}$$

$$= F(\mathbf{x}(t_0))\mathbf{f}(\mathbf{x}(t_0))\ .$$

Hence

$$\mathbf{x}(t) \approx \mathbf{x}(t_0) + \tau[I + \tfrac{1}{2}\tau F(\mathbf{x}(t_0))]\mathbf{f}(\mathbf{x}(t_0))\ .$$

Using this approach we make the following definitions for propagating the state estimate between t_k and $t_{k+1} = t_k + \tau$.

State estimate propagation (forward)

$$\hat{\mathbf{x}}_{k+\frac{1}{2}} = \hat{\mathbf{x}}_k + \frac{\tau}{2}[I + \frac{\tau}{4} F(\hat{\mathbf{x}}_k)]\mathbf{f}(\hat{\mathbf{x}}_k)$$

$$\hat{\mathbf{x}}_{k+1}(-) = \hat{\mathbf{x}}_{k+\frac{1}{2}} + \frac{\tau}{2} [I + \frac{\tau}{4} F(\hat{\mathbf{x}}_k)]\mathbf{f}(\hat{\mathbf{x}}_k)$$

$$k = 0,1, \cdots, N-1$$

$$\hat{\mathbf{x}}_0 \text{ given }.$$

State estimate propagation (backward)

$$\hat{\mathbf{y}}_{k-1}(+) = \hat{\mathbf{y}}_k - \tau[I - \tfrac{1}{2}\tau F(\hat{\mathbf{y}}_k)]\mathbf{f}(\hat{\mathbf{y}}_k)$$

$$k = N,N-1, \cdots ,2$$

$$\hat{\mathbf{y}}_N \equiv \hat{\mathbf{x}}_N$$

The two step procedure for the forward state estimate propagation provides us with a "midpoint" state estimate. We use this to approximate more accurately the one-sided Green's function matrix needed for propagating the covariance matrices. From Section 2 we have the following formulas (for $k = 0,1, \cdots ,N-1$).

Covariance propagation (forward)

$$P_{k+1}(-) = W_k(\tau)P_k W_k'(\tau) + S_k$$

Inverse covariance propagation (forward)

$$P_{k+1}^{-1}(-) = [W_k(\tau)P_k W_k'(\tau) + S_k]^{-1}$$

Covariance propagation (backward)

$$Q_{k-1}(+) = W_k^{-1}(\tau)(Q_k + S_k)W_k'^{-1}(\tau)$$

Inverse covariance propagation (backward)

$$Q_{k-1}^{-1}(+) = W_k'(\tau)(Q_k + S_k)^{-1}W_k(\tau)$$

For the one-sided Green's function matrix we use

$$W_k(\tau) = \sum_{j=0}^{J} \frac{(\tau F_k)^j}{j!}$$

where

$$F_k = \frac{d\mathbf{f}(\mathbf{x}(t))}{d\mathbf{x}'(t)}\bigg|_{\mathbf{x}(t) = \hat{\mathbf{x}}_{k+\frac{1}{2}}} .$$

The choice of J in the above series is determined by the accuracy with which the covariances must be propagated. If we are only filtering, then $J = 1$ is usually sufficient. If we are propagating inverses or the smoother's covariance matrix (see below), then a larger value will be required. The larger value is needed to ensure the consistency of the computed covariance matrix and its inverse. For inverse covariance propagation we use

$$W_k^{-1}(\tau) \equiv W_k(-\tau) = \sum_{j=0}^{J} \frac{(-\tau F_k)^j}{j!} .$$

We also use the same F_k's for propagating the inverses, both forward and backward.

Since J is finite, and because of the limited precision of the computer, some inconsistencies do eventually accumulate between the covariance matrices and their inverses. These inconsistencies may cause the smoother's covariance matrix to lose its positive definite character. The method of Hotelling matrix iteration may be used to bound these inconsistencies [11].

The use of Hotelling matrix iteration is as follows: Let A and B_0, an estimate of A^{-1}, both be given. Then if $I - AB_0$ has spectral radius less than unity, the sequence

$$B_{i+1} = B_i + B_i(I - AB_i) \quad , \quad i = 0, 1, \cdots$$

converges in norm and $\lim_{i \to \infty} B_i = A^{-1}$. Thus if we let $A = P_k(-)$ and $B_0 = P_k^{-1}(-)$, we may iterate on $P_k^{-1}(-)$ and improve the consistency of the computed $P_k(-)$ and $P_k^{-1}(-)$. The Hotelling iteration usually converges very rapidly, so only a few iterations are typically necessary. We alert the reader that for smoothing components of the order of the earth's radius (6,378 km), an error of 10^{-6} in $I - P_k(-)P_k^{-1}(-)$ can imply a spurious estimation error having the order of meters. Thus, if a radar's range measurement is accurate to meters, smoothing annihilates the benefit of filtering, if the accumulation of the propagation errors cannot be bounded.

We observe that a matrix inversion still must be performed in the inverse covariance propagation equations because of the presence of S_k. If S_k is diagonal, this inversion is trivial if the reinforcement method of matrix inversion is employed [11, 13]. This method is valid even if S_k is not diagonal.

The use of the reinforcement method is as follows: Let A be a nonsingular matrix with A^{-1} known. Let B be of positive rank p and assume also that $A + B$ is nonsingular. Express B as the sum of p matrices, each of rank one, say

$$B = B_1 + B_2 + \cdots + B_p \ .$$

If B is diagonal, the above decomposition is trivial. Now let $C_0 = A^{-1}$ and form the sequence $C_0, C_1, \cdots, C_p$ where

$$C_i = C_{i-1} - \frac{1}{1 + \operatorname{tr} C_{i-1} B_i} C_{i-1} B_i C_{i-1} \ , \quad i = 1, 2, \cdots, p \ .$$

Then

$$C_p = (A + B)^{-1} \ .$$

(There are mathematical pitfalls involved in applying this method, see [35].)

The formulas for updating the state estimates and covariance matrices were given in Chapter I. We include them here for completeness.

State estimate update (forward)

$$\hat{\mathbf{x}}_k = \hat{\mathbf{x}}_k(-) + K_k[\mathbf{Z}_k - \hat{\mathbf{x}}_k(-)]$$

Covariance update (forward)

$$P_k = [I - K_k H_k]P_k(-)$$

Inverse covariance update (forward)

$$P_k^{-1} = P_k^{-1}(-) + H_k' R_k^{-1} H_k$$

State estimate update (backward)

$$\hat{\mathbf{y}}_k = \hat{\mathbf{y}}_k(+) + L_k[\mathbf{Z}_k - \hat{\mathbf{y}}_k(+)]$$

Covariance update (backward)

$$Q_k = [I - L_k H_k] Q_k(+)$$

Inverse covariance matrix update (backward)

$$Q_k^{-1} = Q_k^{-1}(+) + H_k' R_k^{-1} H_k$$

In the above equations, the gain matrices are given by

Gain matrix (forward)

$$K_k = P_k(-)H_k'[H_k P_k(-)H_k' + R_k]^{-1}$$

Gain matrix (backward)

$$L_k = Q_k(+)H_k'[H_k' Q_k(+)H_k + R_k]^{-1}$$

Also we have

$$H_k = \frac{d\mathbf{h}_k(\mathbf{x}(t))}{d\mathbf{x}'(t)}\bigg|_{\mathbf{x}(t) = \hat{\mathbf{x}}_k(-)},$$

and R_k is the measurement noise covariance matrix associated with $\mathbf{Z}_k$, $k = 1, 2, \cdots, N$.

The purpose of using the same reference vector for evaluating H_k is to ensure the consistency of the covariance matrices and their inverses. If we were, for example, to evaluate H_k with respect to $\hat{\mathbf{x}}_k(-)$ for the forward filter and with respect to $\hat{\mathbf{y}}_k(+)$ for the backward filter, we would have violated several assumptions used in deriving the smoother equations [such as (1) of Section 3].

In preparation for smoothing we must then store the midpoint estimates, $\hat{\mathbf{x}}_{k+\frac{1}{2}}(-)$, for covariance propagation, and the endpoint estimates, $\hat{\mathbf{x}}_k(-)$, for updating. After completing the forward filter's pass through the data, we have $\hat{\mathbf{x}}_N$ and P_N at t_N. These provide the initial conditions for the backward filter and smoother propagation equations,

$$\hat{\mathbf{y}}_N = \hat{\mathbf{x}}_N$$

$$M_N = P_N .$$

Also, we require

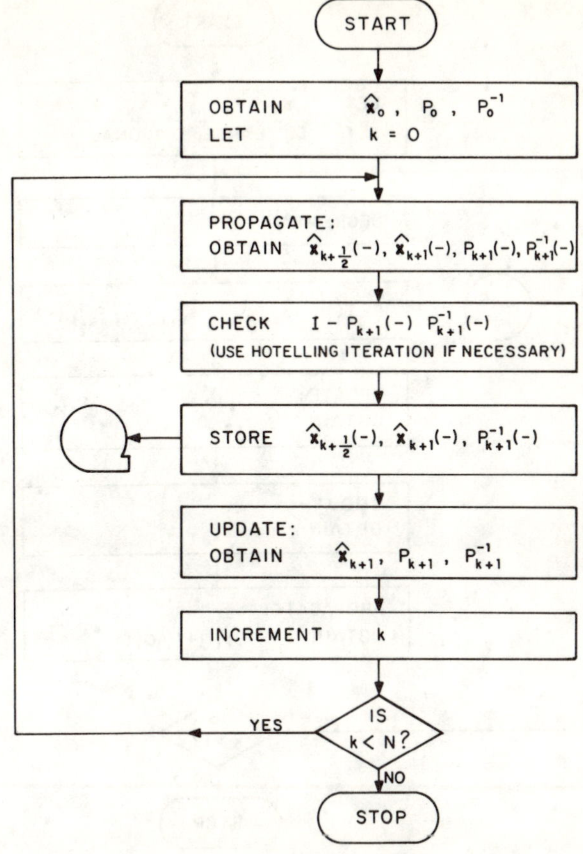

Figure 2
Forward Filter for Smoothing

$$Q_N^{-1} = 0 \ .$$

The requirement that Q_N^{-1} be zero obviates the use of the D-form of the RTS smoother equations, since both Q and Q^{-1} are required. For practical backward filtering we need Q to be nonsingular for updating the state estimate. Thus the forms of the smoother propagation equations we use require either P^{-1}, P, M or Q^{-1}, M. We list the more useful equations from Section 3.

Smoother propagation (P-form)

$$M_k = W_k(-\tau)\{S_k - S_k P_{k+1}^{-1}(-)S_k$$

$$+ [S_k P_{k+1}^{-1}(-) - I]M_{k+1}[S_k P_{k+1}^{-1}(-) - I]'\}W'(-\tau)$$

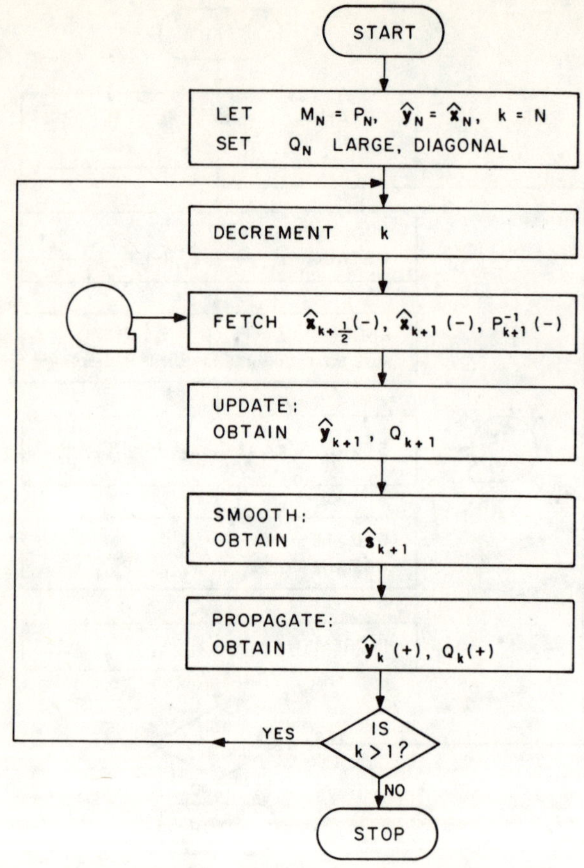

Figure 3
Backward Filter for Smoother (P-form)

Smoother propagation (Q-form)

$$M_k = W_k(-\tau)[-S_k - S_k Q_{k+1}^{-1} S_k + (S_k Q_{k+1}^{-1} + I)M_{k+1}(S_k Q_{k+1}^{-1} + I)']W_k'(-\tau)$$

Smoothed state estimate (P-form)

$$\hat{\mathbf{s}}_k = \hat{\mathbf{y}}_k + M_k P_k^{-1}(-)[\hat{\mathbf{x}}_k(-) - \hat{\mathbf{y}}_k]$$

Smoothed state estimate (Q-form)

$$\hat{\mathbf{s}}_k = \hat{\mathbf{x}}_k(-) + M_k Q_k^{-1}[\hat{\mathbf{y}}_k - \hat{\mathbf{x}}_k(-)]$$

The overall forward-backward filter smoother procedure is illustrated in Figs. 2 and 3.

5. A NONTRIVIAL SMOOTHER EXAMPLE

As an application of the analyses presented above, we consider the problem of determining, from radar observations, the motion of a three degrees of freedom aerodynamic reentry vehicle. The system and measurement models are those described in Section 1. For purposes of evaluating estimation accuracy, a set of simulated data was generated by integrating the system model differential equation with a sixth order Gear predictor/corrector integrator [19]. The initial conditions corresponded, approximately, to a vehicle at 100 km altitude and having a velocity relative to earth of about 6.8 km/sec. A synthetic ballistic coefficient profile was used to simulate drag and also included effects such as free molecular flow, slip, and laminar regimes, as well as boundary layer transition to turbulent flow.

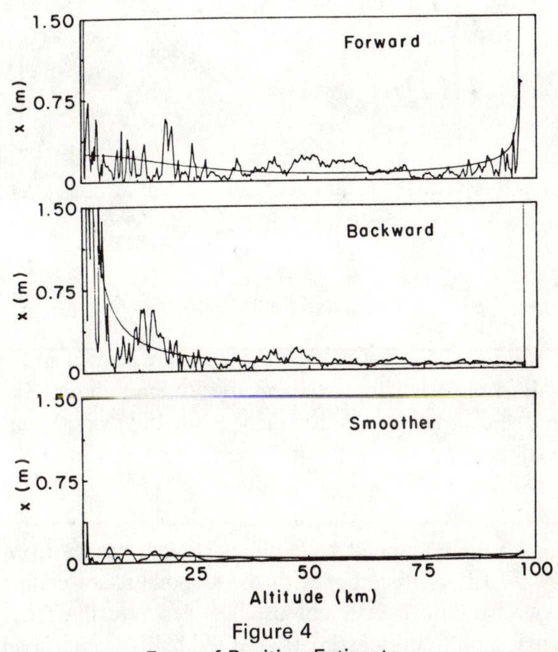

Figure 4
Error of Position Estimate

Radar observations were generated by sampling the integrated trajectory at a 10 Hz rate. The radar data were perturbed using Gaussian noise with standard deviations of 0.2 m in range and 0.005° in angle. The radar was positioned on the earth near the equator, such that the initial state vector corresponded to radar coordinates of 150 km range, 63° azimuth, and 28° elevation. The impact point was chosen near the radar, and reentry flight time lasted approximately 25 sec.

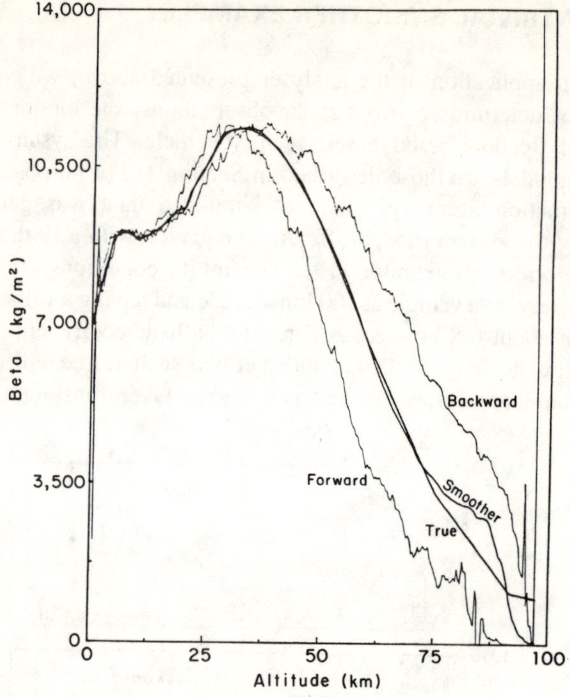

Figure 5
Estimated Ballistic Coefficient

The results of the estimation example are shown in Figs. 4 and 5. Figure 4 illustrates the improvement due to smoothing. There the noisy curves are the estimation position error while the smooth curves represent the associated standard deviation, computed from the error covariance matrices.

In Fig. 5 the estimated ballistic coefficient curves from the forward and backward filters, together with the smoother, are compared with the actual (true) ballistic coefficient used to generate the data. The curves illustrate that high fidelity estimation of the ballistic coefficient is possible using only position data (range, azimuth, and elevation). The minor departure of the smoother from the true curve at high altitude is an indication that the atmosphere is extremely thin there, and the ballistic coefficient starts to lose its observability as altitude is increased.

References

[1] V. I. Arnold, *Ordinary Differential Equations*, The M. I. T. Press, Cambridge, MA, 1980

[2] J. W. Austin and C. T. Leondes, *Statistically linearized estimation of reentry trajectories*, IEEE Trans. on Aerospace and Electronic Systems, *AES-17*, No. 1, 54–61, (1981)

[3] F. N. Barnes, *Stable member equations of motion for a three-axis gyro stabilized platform*, IEEE Trans. on Aerospace and Electronic Systems, *AES-7*, No. 5, 830–842, (1970)

[4] G. A. Blass, *Theoretical Physics*, Appleton-Century-Crofts, NY, 1962

[5] D. A. Castanon, *Reverse-time diffusion processes*, IEEE Trans. on Inform. Theory, *IT-28*, No. 6, 953–956, (1982)

[6] C. B. Chang, R. H. Whiting, and M. Athans, *On the state and parameter estimation for maneuvering reentry vehicles*, IEEE Trans. on Automatic Control, *AC-22*, No. 1, 99–105, (1977)

[7] E. A. Coddington and N. Levinson, *Theory of Ordinary Differential Equations*, McGraw-Hill, NY, 1955

[8] J. A. D'Appolito, *A simple algorithm for discretizing linear stationary continuous time systems*, Proc. IEEE, *54*, No. 12, 2010–2011, (1966)

[9] G. Dahlquist and Å. Björck, *Numerical Methods*, Prentice-Hall, Englewood Cliffs, NJ, 1974

[10] R. Deutsch, *Orbital Dynamics of Space Vehicles*, Prentice-Hall, Englewood Cliffs, NJ, 1963

[11] D. S. Dwyer and F. V. Waugh, *On errors in matrix inversion*, J. Amer. Statist. Assoc., *48*, 289–319, (1953)

[12] P. R. Escobal, *Methods of Orbit Determination*, R. E. Krieger Pub. Co., Huntington, NY, 1976

[13] D. K. Faddeev and V. N. Faddevva, *Computational Methods of Linear Algebra*, Freeman, San Francisco, CA, 1963

[14] D. D. Falconer and L. Ljung, *Application of fast Kalman estimation to adaptive equalization*, IEEE Trans. on Commun., *COM-26*, No. 10, 1439–1446, (1978)

[15] J. L. Farrell, *Integrated Aircraft Navigation*, Academic Press, NY, 1976

[16] R. J. Fitzgerald, *Divergence of the Kalman filter*, IEEE Trans. on Automatic Control, *AC-16*, No. 6, 736–747, (1971)

[17] D. C. Fraser and J. E. Potter, *The optimum linear smoother as a combination of two optimum linear filters*, IEEE Trans. on Automatic Control, *AC-14*, No. 4, 387–390, (1969)

[18] B. Friedland, *Analysis of strapdown navigation using quaternions*, IEEE Trans. on Aerospace and Electronic Systems, *AES-14*, No. 5, 764–768, (1978)

[19] C. W. Gear, *Numerical Initial Value Problems in Ordinary Differential Equations*, Prentice-Hall, Englewood Cliffs, NJ, 1971

[20] A. Gelb, *Applied Optimal Estimation*, The M. I. T. Press, Cambridge, MA, 1974

[21] H. Goldstein, *Classical Mechanics*, Addison-Wesley, Reading, MA, 1959

[22] P. R. Halmos, *Finite-Dimensional Vector Spaces*, D. van Nostrand, Princeton, NJ, 1958

[23] F. B. Hildebrand, *Introduction to Numerical Analysis*, McGraw-Hill, NY, 1956

[24] A. H. Jazwinski, *Stochastic Processes and Filtering Theory*, Academic Press, NY, 1970

[25] P. G. Kaminski, A. E. Bryson, Jr., and S. F. Schmidt, *Discrete square root filtering: A survey of current techniques*, IEEE Trans. on Automatic Control, *AC-16*, No. 6, 727–735, (1971)

[26] C. T. Leondes, J. B. Peller, and E. B. Stear, *Nonlinear smoothing theory*, IEEE Trans. on System Science and Cybernetics, *SSC-6*, No. 1, 63–71, (1970)

[27] D. M. Leskiw and K. S. Miller, *A comparison of some Kalman estimators*, IEEE Trans. on Inform. Theory, *IT-25*, No. 4, 491–495, (1979)

[28] D. M. Leskiw and K. S. Miller, *Convergence of polynomial least squares estimators*, Proc. IEEE, *70*, No. 5, 520–522, (1982)

[29] D. M. Leskiw and K. S. Miller, *Convergence properties in Kalman filtering*, Journal Inform. and Optimization Sci., *1*, 197–213, (1980)

[30] P. B. Liebelt, *An Introduction to Optimal Estimation*, Addison-Wesley, Reading, MA, 1967

[31] S. W. McCuskey, *Introduction to Celestial Mechanics*, Addison-Wesley, Reading, MA, 1963

[32] D. W. Marquardt, *Generalized inverses, ridge regression, biased linear estimation, and nonlinear estimation*, Technometrics, *12*, No. 3, 591–612, (1970)

[33] J. J. Martin, *Atmospheric Reentry*, Prentice-Hall, Englewood Cliffs, NJ, 1966

[34] K. S. Miller, *Linear Differential Equations in the Real Domain*, W. W. Norton, NY, 1963

[35] K. S. Miller, *On the inverse of the sum of matrices*, Math. Mag., *54*, No. 2, 67–72, (1981)

[36] K. S. Miller, *Some Eclectic Matrix Theory*, R. E. Krieger Pub. Co., Malabar, FL, 1986

[37] K. S. Miller, *Vector Stochastic Processes*, R. E. Krieger Pub. Co., Huntington, NY. 1980

[38] K. S. Miller and D. M. Leskiw, *Nonlinear estimation with radar observations*, IEEE Trans. on Aerospace and Electronic Systems, *AES-18*, No. 2, 192–200, (1982)

[39] F. J. Murray and K. S. Miller, *Existence Theorems for Ordinary Differential Equations*, R. E. Krieger Pub. Co., Huntington, NY, 1976

[40] C. F. O'Donnell, *Inertial Navigation*, McGraw-Hill, NY, 1964

[41] H. E. Rauch, F. Tung, and C. T. Striebel, *Maximum likelihood estimates of linear dynamic systems*, AIAA Journal, *3*, No. 8, 1445–1450, (1965)

[42] E. J. Routh, *Dynamics of a System of Rigid Bodies*, Dover Publications, NY, 1960

[43] A. K. Sen, *Tools of kinematics: A critical review*, IEEE Trans. on Aerospace and Electronic Systems, *AES-15*, No. 1, 40–46, (1979)

[44] M. I. Skolnik, *Introduction to Radar Systems*, McGraw-Hill, NY, 1980

[45] T. T. Soong, *Random Differential Equations in Science and Engineering*, Academic Press, NY, 1973
[46] J. L. Synge and B. A. Griffith, *Principles of Mechanics*, McGraw-Hill, NY, 1942
[47] R. W. Truitt, *Hypersonic Aerodynamics*, Ronald Press, NY, 1959
[48] R. von Mises, *Theory of Flight*, McGraw-Hill, NY, 1945
[49] E. Wong, *Stochastic Processes in Information and Dynamical Systems*, McGraw-Hill, NY, 1971

16 S²